DREAMING THE PAST, DREAMING THE FUTURE.

A *Herstory* *of the Earth*

DIANE STEIN

The Crossing Press
Freedom, California 95109

I would like to thank the women who gave so much time and energy to the forming of this book. First, the channelers: Mari Aleva, Marion Webb-Former, Tanith, and Laurel Steinhice, who did private readings and answered so many questions. I would also like to thank the following for their encouragement through the writing process: Caridwyn and Mari Aleva, Elisa Robyn, Eileen Sullivan and David Speer. For help in editing and their thoughtful comments on the manuscript, I thank Nett Hart, Tanith, and Caridwyn Aleva.

Without the help of these patient, loving people, this book could not have happened. Nor could it have happened without the faith, expertise and patience of the folks at The Crossing Press. I thank you more than I can say.

Cover art by Sudie Rakusin
Cover design by Tom Trujillo

Printed in the U.S.A.

ISBN 0-89594-519-3 (pbk.)
ISBN 0-89594-520-7 (cloth)

Contents

When you heal yourself and assist others with their self-healing, you heal the earth. For the earth is one with everyone and every creature. All are part of the earth. Those that creep and crawl and go upon many legs, those that fly in the air, those that go upon four legs or two, are the Earth Mother's children. All are part of the earth. The healing of any one of these is healing that contributes to the whole.

You do *make a difference. That is the message we bring.* It does matter. *The simple loving of the earth is healing. When you love the earth you bring energy for her healing. There is interconnectedness between earth energy and personal energy, and use of that energy is what the transition is all about.*

—Laurel Steinhice channeling the Earth Mother,
Earth Changes Channeling, June 3, 1990, p. 1.
See Appendix.

A Meditation

In the quiet of the end of the day, alone or with other women, prepare a ritual space. Take a shower or a cleansing bath, take the phone off the hook, and come to the altar skyclad or wearing loose, natural fabric clothing. Light one or more candles and incense if you choose, shut the electric lights off, purify, and cast a circle of protection. Salute the earth and sky, and invite into the circle Goddesses of the four directions (or six, or eight) and the center. Invite Goddesses of peace and healing, of growth and inner change: Yemaya, Isis, Kwan Yin, Tonantzin, Demeter, White Buffalo Calf Woman. On a Full Moon, draw down the Goddess' power, invoking her within yourself. On the night of a Sabbat, invite into the circle Mother Earth.[1]

Now do this meditation, seated or lying down within the circle:

Relax every part of your body, moving from toes to head. Tense and relax each set of muscles, repeating the sequence a second time if needed. Then, imagine yourself walking through a door in the altar to the outside. Follow a path through a green, wooded forest to the seashore. Hear and see the tree leaves rustling a welcome, feel a gentle wind, smell the green of the living woods, greet a squirrel or a rabbit that walks beside you for awhile. Thank the Goddesses of forests, Artemis perhaps, for the woods.

Come to a beautiful beach and climb over white-sided sand dunes to reach the shore. The sky is blue, the sun gently shining; Yemaya's foamy waves rush in and out along the sand. Come to the edge of the water. Look out at the waves and then up at the sky. Discover that you have wings. Flex them and examine their beauty; spread them out and fly.

Rise from the sea sand into the air, into the blue of the bright sky. Rise higher and higher above the earth, above the fleecy clouds. The sky grows bluer and then blue-black as you rise. Watch the stars come out, thousands of them in their constellations, and see the Milky Way. Many Goddesses have their stories in the stars, and the Milky Way is the breast milk of Isis and of Spider woman. See the moon and the earth rise above her, and remember the many moon Goddesses. Fly further, above the earth and moon, until you can look down

v

upon them in their beauty in the universe below.

Circle the earth and moon, flying in an orbit that includes them both. Look down at life on earth, and really look at it. With each orbit above the planet, take notice of something that needs healing. Notice the choking pollution, mountains of garbage, the greed of corporations and governments, the many wars, the hate. Notice the Alaska oil spill. Notice China and Africa, apartheid, violence and oppression in too many nations — including "free" ones in the West. Notice the poverty of women and children everywhere, the misogyny. See the abuse of animals and the species becoming extinct — so many of them — the dying dolphins and the poisoning of songbirds. See governments, politicians, sterile patriarchal religions; see racism, sexism, nationalism, ethnicism. Notice AIDS, heart dis-ease and cancer, environmental illness, gay bashing, incest, rape, ablism and ageism. Notice the lack of medical care for too many and the abuse of healing by technology and greed. Notice how many are hungry, homeless, unloved and unvalued. Notice patriarchy, the suppression of the feminine, and the pain of most of the planet's people.

Now circling again, notice all the things that need healing in yourself: childhood abuse or incest, growing up female where a male was wanted, growing up poor, or Black or disabled. Notice how patriarchy has oppressed your life, restricted your choices, kept you in the ghetto or homeless, or with too many pregnancies and no means to raise the children, kept you under-educated, under-employed or unfree. Notice the things that are needed in your life now: opportunity, inner power, prosperity, choices, healing, peace of mind and love. Notice how you have overcome and what remains to do. Honor yourself and your strength.

Circle again, this time taking each thing that needs healing in yourself and filling its image with light. Envision the hurt, then the solution, and imagine the solution put into effect. The solutions can be possible or not, can be magick, but always harming none. See the hurt healed, and what it would be like. Feel how it would be, see it, experience it in all the senses, live it. Go on to the next hurt and heal that one in the same way. And the next, until all of the issues are solved. Experience what life would be like if all of the pain were healed, if everything needed were there and available. Imagine life prosperous and loved, filled with excitement and joy, opportunity and choice, filled with contentment, security and peace. Circle the earth again, living this feeling of wholeness. Thank the Goddess and yourself for creating it.

Now look back to the planet and do the same thing for all of the wrongs of the earth. With each flown circle, fill an issue with light, heal it, experience it healed, then go on to the next healing. With all of the wrongs healed, the earth is filled with brightness. She is clean and unpolluted, her people are safe and free. Everyone is sheltered, fed, clothed and taken care of, everyone is respected, including the animals and planet. The dis-eases are gone, the isms and the patriarchy. Women are honored and the feminine is valued. The Goddess has returned. Imagine the beauty of the healed earth. What would it be like to live there? You and the Goddess have created it; thank Her and yourself.

Make one more orbit around the earth and moon and then fly back to the beach. See the planet healed and whole, and yourself healed and whole living upon it. Remember what it would be like, and land on the beach with the memories intact. Fold your wings, and rinse your feet in the waves at the ocean's edge. Thank Yemaya, the ocean, and climb over the sand dunes to the forest path. A rabbit meets you under an oak tree and the birds sing. Follow the path to the door in the altar, thanking Artemis of the forest before going through the door. Open the door and return to your altar, to your self sitting on the floor before it. Remember again the vision of a healed world and a healed self. Bring it back with you to remember always, as you come back to now.

Stretch and move slowly coming back to the earthplane. Thank the Goddesses of the directions, ask for their strength in daily life, open the circle. Put out the incense and blow out the candles. Keep the image of a healed life and a healed world before you; you have created them, now put them into Be-ing. For the good of all.

Notes

1. For a full book on ritual, see Diane Stein, *Casting the Circle: A Women's Book of Ritual* (Freedom, CA: The Crossing Press, 1990).

Introduction

What Are the Earth Changes?

In the late twentieth century there is a growing awareness that we are doomed as a species and planet unless we have a radical change of consciousness. The re-emergence of the Goddess is becoming the symbol and metaphor for this transformation of culture.

Elinor W. Gadon[1]

The popular New Age movement envisions the future as a world of peace, a world far different from what we live in now. Their New Age is a utopia, a word coined by Plato that means "no place," but a concept that describes all wishes come true. To live in utopia — or the New Age, Age of Aquarius — means to live in paradise. To live in the New Age means to live on a planet where there are no wars, where everyone is "enlightened," and where everyone has happiness and well-being. How that world is to happen, or what else other than peace it really and specifically means, is generally undefined. In the literature and among New Age lecturers and teachers, the utopia of the New Age is pretty much what the individual wants to make it.

When pressed for details of what this New Age will be, most versions are male-oriented, white-skinned, suburban middle class, American, and of course heterosexual. They are a streamlined "things as they are," complete with machines to do all the work, longer lives in always youthful bodies, and even immortality. Rarely does a vision of this coming New Age include other races than white, other cultures than American or at least Western, or a point of view that is not patriarchal. While the New Age male prides himself on his sensitivity, he still has all the aces in his hand. It's his world, or he thinks it is, and women's role is still the same.

Since the start of the Industrial Revolution, patriarchy's potential for harm has taken on life-threatening global proportions. Along with the rape of the natural resources of the planet, industrialization meant an intensified rape of the feminine. Women became objects to be used and owned, managed through poverty, and the earth to be used and owned along with them. The prevailing practice was to dominate the earth and all that lives on it, including women and

1

any but white males, with no regard for life or the future. Women continue to be cheap labor in the home and factory in every industrialized nation, despite the recent advances. Men are often just as exploited, while lacking the awareness to know. The earth is something to tear apart for profit, leaving waste and desolation as a byproduct, with money the only value scale. As technology increased, so did the damage.

While this austere pillaging has been softened in the New Age, particularly under the influence of the women's movement and the clear and present danger to the environment, there is no new vision beyond (or reaching before) the patriarchal model. The concept of "no more war" is indeed utopian, but the New Age had no real definition of what utopia is, no ideas of what to aim for in a reclaimed world. It has no concept to change the root of the problem — the patriarchal mindset. Without that change, "no more war" is impossible.

Coming out of the 1960s with the New Age, however, are the feminist and Goddess movements, and women have a much clearer idea of what a New Age could be. Women's vision is not a cleaned up "things as they are" but a radical new way of re-ordering the structures of the world. The feminist movement began with political theory, herstory and literature and evolved into a reclaiming of women's Be-ing in the world and of the world itself. To bring about "no more war" in a women's New Age, the honoring of the female is primary. It includes the revaluing of women, of the earth and of all living things.

A women's utopia is something to work for and create, rather than to wish for as "out there" in a someday future. The revaluing of the feminine means a revamping of governments and schools, of the monetary and trade system, of world politics, of wealth and food distribution, and of religion. It means cooperation and mutual respect instead of competition and aggression. Women have thoughtful ideas of what is ideal and a willingness to learn ways to make their ideals workable. A New Age in women's movement thinking is multi-racial and multicultural, honoring the diversity of all people. It begins with placing life first by taking care of the earth and of all its inhabitants — plant, animal or human. It means an end to tolerance for patriarchal violence, greed, misogyny, manipulation and money-first non-ethics. It means giving women the total right to choose the uses of their bodies and lives, and of placing women in at least full equality. It means having respect for all people, and instituting (and enforcing if necessary) values that place the good of all as primary, not only the good of the rich or those in power. The central symbol of women and the earth in that New Age is the Goddess.

Why has the Goddess become the symbol for a women's New Age and the basis for women's utopian theory? While the patriarchal god places divinity somewhere in heaven, the Goddess is the earth herself. She is the earth, moon, universe and stars, and exists in everything that lives, from a blade of grass to an atomic scientist. Taking care of the earth, and of each other, is taking care of

Goddess — who is the living part of her celebrants and exists with them. The concept of a divinity that is part of everything is called immanence, and is a central part of feminist wicce or Women's Spirituality, the Goddess religion. Notice too that the Goddess is female, and how that changes women's views of divinity and the world.

The earth as a Goddess has the name of Gaea, who was the pre-Hellenic Greek earth mother. A number of early cultures saw the planet as a Goddess and had names for her, but Gaea (sometimes spelled Gaia) is the name that had stuck in the English-speaking West. To celebrate the Goddess, as Gaea or any other of her many names, is to view the planet as a living Be-ing, as the birth mother of all. Like any human or animal mother, she produces life from her body by giving birth, and nourishes it from her breasts once it is born. There is no birth or survival after birth without her. No celebrant of Gaea would dream of strip-mining or of loading the earth's body with garbage or toxic wastes. It is obviously against the best interests of those wanting the Mother's bearing and nourishment, of those wanting to stay alive.

The patriarchal god is not immanent. He exists somewhere off the planet, in a cloud in heaven, and is totally apart from those he created. Male creation is a fantasy. Any adult and most gradeschool children know that males have no capacity for giving birth or creating life. But patriarchal religions inform us that god made Adam and Lilith from dust and Eve from Adam's rib. This was not credible to the Goddess farming cultures of the time, and not conducive to life-valuing thinking. God told Adam to multiply and dominate the earth, and his patriarchies have done just that, to the near extinction of life on the planet. They have had to enslave the Goddess, and the life-giving, life-nurturing women in her image, to do it.

Where god is female, is Goddess, it's a different story. The early matriarchal, lunar/Goddess-based societies lived in harmony with the cycles of the earth. They "walked gently on the Earth Mother" as the Native Americans put it. Virtually all nonwestern cultures began as Goddess cultures, adopting the patriarchy and male god only recently and unwillingly. From about 15,000 BCE to 2,400 BCE northern patriarchal tribes migrated south through Europe and north Africa, decimating the matriarchies and instituting the current order. The same type of invasions seem to have happened in South America as well, before the much later (16th century) Spanish conquest of native North and South America. In China, the legendary Era of Great Purity ended about 4,000 BCE; it had been a Goddess matriarchy. With the invasions came the end of world peace and the supression of the Goddess and of women.

A New Age for women can be a returning of the earliest age, a returning to Goddess/women's values and a way of recreating a world that honors the earth and operates for the good of all. Such a New Age would begin with valuing women as images of the life-giving Gaea. With perceiving the planet as our mother, a women's New Age would "walk softly on her" again. There would be no more toxic or nuclear wastes, polluted streams or oceans, noxiously poisoned air, destroyed ozone layer, or leached and chemically deadened soils. A simple recognition of the earth as life is the beginning of major changes in the environment.

Honoring women and the planet also has clear implications for honoring all that lives. The current order of racism, nationalism and ethnicism would be unthinkable in a world where life is venerated. All that lives is honored, and is honorable. No one color of skin, or sexual orientation, or gender, or national origin, or culture, or body size or age could be declared any more or less desirable than any other. The life of each individual is valued, and no life is more or less worthy than any other life. The life in feathers, fur or fins is also life and has the same worth and right to be here. Nor is the life of plants and trees, so necessary now for planetary survival, to be discounted or devalued. Nothing that lives is here to be dominated or exploited — all are Goddess. The New Age utopia of "no more war" would become reality.

A Goddess New Age means a returning to values of living on the planet, on Gaea, in harmony and respect. The first and primary law of the Goddess religion is "harm none." Seldom today does a factory owner of shopping center builder consider the effects his establishment has on the rest of life. Does his installation tear down forest or rangeland, or the habitat of animals and birds? Do the activities of his establishment pollute the water table or the air? How badly and what is harmed? Can he control or prevent the damage? Is his factory or shopping mall ecological — as they say in neurolinguistic programming vocabulary — is it good for all of life? Until there can be a "yes" to that question, the factory or shopping center in a Goddess New Age would not be built. It must harm none and be for the good of all.

The same is true for individuals. Does the best qualified individual get the job, or does it go only to a white male? Is the job, or the training for it, denied to women, to lesbians or gay men, to disabled people, to people with AIDS, to Black or Chicana or Native American or Asian people? Is such a denial good for all of life, is it ecological, or does it affront Gaea who is the life force/earth? When something harms none, it is good for all of life and benefits all. What is good for anyone is good for everyone, including the planet and environment.

And what about governments? Under a Goddess New Age, could the Equal Rights Amendment continue to be defeated? Could any bill assuring equal civil rights — to lesbians and gays, to women, to disabled people, to the elderly — cause the huge uproars in Congress that such bills cause today? Would women have to fight so desperately for reproductive freedom? For health care, Aid to Dependent Children, adequate food, jobs, shelter or daycare? Could apartheid continue to exist in South Africa or the suppression of women in the Middle East? Could native peoples be forced from their lands in North or

South America, or forced into governments that they don't want to live under? Would China be permitted to slaughter thousands of students demonstrating for democracy?

In a women's New Age, government would be very different from the undefined utopia of the popular New Age movement. Of course there will be global peace and an end to war, both planetary and local, but that's only the beginning. Government will be multinational and based on the law of "harm none," the natural law of the Goddess. There will be cooperation among nations for the good of all. Economies will be based on the needs of the world, not just in the needs of a few rich patriarchs. People will be taken care of, and food, shelter, clothing, heat in cold climates, and medical care will be considered human rights. Women will have freedom of choice for reproduction, and decreasing population will be a goal. Famine in one part of the world will be offset by the surpluses from other places. The emphasis of government will no longer be on defense spending, but on research to end AIDS, cancer and heart dis-ease and to find nonpolluting fuels. The respected, cleaned up environment will produce healthier food and healthier people. And food will no longer be fast foods but healthy, organic, unchemicalized crops.

Rather than pushing a button to take care of any work, a women's New Age will go back to older ways. There might be more horses than automobiles again, or a new form of nonpolluting, nondepleting fuel. There might be more farmers than factories. The days of industrialization for its own sake will change radically, or decline. The Western patriarchal dependence on gadgets and over-mechanization will be replaced with a way of life that is closer to the earth and in harmony with her, and that "harms none." Electricity will be generated by water, wind or solar power, or by magnetics rather than by fossil or nuclear fuels. There will be less of it, consumed for more important needs that it is now. Life will be more fulfilling, more fully lived, and more authentic, natural, real.

Since women are the architects of the Goddess New Age, it is up to us to create and implement the utopias, and to learn from the mistakes of the present day. A women's New Age will be an age of equality, an age of the valuing of women's culture, of the skills that developed civilization. Without the women's-work skills of food gathering and growing, pottery and clothmaking, homebuilding, metalwork and tool-making, basketry, animal care, midwifery and healing there would be no civilization at all. Such skills will regain importance in women's utopian New Age; they have been long devalued in this one. It will be more important to grow the healthiest food crop than to profit unethically from a political deal. It will be more important to teach healing and farming than military strategies, more important to learn about other nations and cultures than to fight with them. Cleaning up the planet and keeping it clean will be more important than producing VCRs. How to live gently on Mother Earth, on Gaea, will be taught in schools.

But whatever women can create in a coming New Age, we still have the present, the patriarchal order that is far from utopian to deal with. How will we get from where we are to where we want to be? According to popular New Age thinking, as people become more aware, the old order will crumble and die, and the new will begin. What that new order will be depends on women, but it has to first be able to manifest, and the current world needs to change a great deal for that to happen. Patriarchy isn't going to end easily; the boys aren't about to turn the world over to the women and take a back seat. Patriarchy began with violence against women and the earth and has built its power structure on greed, misogyny and war. How do we get to the utopian New Age?

The records of virtually every ancient culture that have been recovered and deciphered tell the story of major changes of the earth. Every culture has its legends of the great flood, the christian Bible being a relatively recent one. Every culture has its legends of lost continents, civilizations that were techno-logically advanced and that were destroyed by earthquakes and polar shifts, by wars or by misuse of technology. Science has evidence of earth changes of the very ancient past, but still refuses to credit the evidence as more than myth. To do so would upset the teachings of the Bible and the foundations of scientific thought. Yet, the remains of tropical animals have been found quick-frozen for thousands of years, showing incredibly rapid freezing and climatic shifts. Huge dinosaurs were once a primary lifeform and they disappeared almost overnight. Ancient artifacts believed to be electric lightbulbs, batteries and computers have been discovered, and their cultures' writings describe visits from other planets. No one knows how the stones for the pyramids or Stonehenge were actually put in place, and whole cities built of such stone blocks lie unexplored under the seas. The *Mahabarata* of ancient India describes what could only be a nuclear blast.

The New Age believes that what once destroyed these lost civilizations is about to happen on earth again. Put in the terms of a battle between good and evil, the theory goes that because our highly technological society has abused the planet and abused all of life, it will be destroyed as the ancient civilizations of Atlantis and Mu were destroyed, and for the same reasons. Some see the earth changes as a major cleansing of the planet, and a cleansing of the human lifeform that has polluted it. Interpretation of ancient records and of the writings of the Bible and Koran are shown as sources for the idea that the ancient past will be repeated. When the destruction is over, the present order will no longer exist, nor will most people remain on the planet. The New Age will be free to rebuild, and to avoid making the same mistakes again. The return of a messiah, the New Age awareness or Christ Consciousness, will go with it. This is the teaching of the popular New Age.

There is a great deal of disguised Bible thumping here and probably a lot of misinterpretation. According to some writers, the current order must be totally destroyed before the new can begin. This makes *some* sense, but does it have to be destroyed violently? If people everywhere — primarily women comprise both the New Age and Goddess movements — are changing in their consciouness from a life-abusing to a life-affirming (patriarchal to matriarchal/Goddess) value system, why can't the earth change in peaceful ways? Why are earth-quakes, volcanos, tidal waves and world wars more earth changing than a new system of thinking that rescues the earth from within its people? Why not cleansing by environmental action instead of a new ice age or a pole shift? Why not use our technology to help instead of to harm, to heal instead of to destroy? Is the concept of earth changes a physical fact or a metaphor?

Perhaps it's women's choice, as the changes in awareness needed to create a new world are changes to Goddess consciousness. In the Goddess religion we have long been taught that what women visualize they create in the material/real world. The bumper-sticker reads, "Visualize World Peace." Why not visualize, and then create in fact, world peace and along with it a clean planet, an end to the abuses against women and against life. The changes, then, begin in the individual, in teaching her to visualize and what to ask for. Remember the wiccan rede, "Be careful what you ask for, you might get it." By asking for a women's New Age, a Goddess age of women's values and respect for all that lives, and then by learning to create it, women can change the world in peaceful, nonvio-lent ways. The older order can indeed crumble without a bang. The earth changes happen from within.

To do this, enough people — men and women — need to know what to ask for. This raising of awareness and consciousness is, in my opinion, the whole point of a theory of earth changes or a women's New Age. In Women's Spirituality, we are taught that our actions and thoughts have consequence, and that women's use of that consequence creates the world as we wish it. In the Goddess movement, women learn who they are, learn about Goddess-within, and learn the women's New Age values that affirm all life. Women who celebrate the Goddess learn to create what they want in their lives, and learn to "harm none." Every decision and choice is made with the consideration of, "Is it ecological, is it good for all life?" In learning to create only those things that *are* ecological, women learn to create the utopian New Age. The learning begins with the law of "harm none" and the reminder, "You are Goddess." It begins with the individual, grows to the small coven or group, and expands to whole communities, nations, planets.

The popular New Age concept of destroying the earth to save it is a product of patriarchal negativity. If enough women actually believe that the planet has to go through violent cleansing, world wars, ice ages, earthquakes and pole shifts, their ability to visualize such occurrences could actually cause them to happen. Thought is the strongest power in the universe, use it wisely. When women don't know that — and it's only one power that the patriarchy conceals from its people — the force is inevitably misused. New Age teachings have advertised the concept of violent, irrevocable destructions on the earth, without presenting a plan or even making the attempt to change it. Not knowing about women's consequence leaves women sitting back and waiting for the worst, hoping to reincarnate in the New Age paradise later. Women's Spirituality teaches about Goddess-within and consequence; women have the power to positively change the world, to create what we want to happen. It's the major thing that women have to teach the mainstream.

The world as we know it is at a crisis point, a choice point of live or die. The patriarchal order has created its image on the planet negatively, dominating and subduing the Goddess Earth until she can almost no longer bear or nourish life. The patriarchy has harmed many and judged good to mean benefit to the powerful few — and those few are never women. The matriarchal Goddess order sees a new way, a way of harming none and benefiting all, male and female, of "walking gently on the Earth Mother."

Slowly the New Age and women's New Age, the wiccan and Goddess and Women's Spirituality movements, have envisioned and begun to create a new way of ordering the world. This new way is very old, very ancient, and is a returning to a past from before the great flood, before earth changes destroyed civilizations that may have been the first Goddess matriarchies. By returning to women's values that have been lost from the ethics of the world for thousands of years, the earth changes, and the transition to a utopian New Age is begun. Beginning from within yourself, visualize these changes gently.

Since the Harmonic Convergence, August 17, 1987, women have begun to feel the effects of the transition from old to new. The date, chosen from the ancient Mayan calendar through the work of José Argüelles in his book *The Mayan Factor* (Bear and Co., 1987), was designated by popular New Age leaders as an opportunity for awakening spiritual awareness in people everywhere. The weekend offered a unique astrological conjunction of planets that also aroused interest, and the concept went beyond New Age and into other alternative movements, including the women's Goddess movement. Millions worldwide participated in rituals, gatherings, sunrise services and meditations — alone or in groups, in public gatherings or at home.

For many women, this was their first exposure to the concept of ending and changing cycles on the planet. It was a point of changing awareness, astrologically and galactically, but also in the mass psychology of being part of the activities and groups. The weekend was defined as, "The point at which the counter-spin of history finally comes to a momentary halt, and the still imperceptible spin of post-history commences."[2] In this energy shifting of the earth, welcomed and magnified by the thought power of the many who participated in it, many women began internal energy shiftings of their own. These took the form of some spectacular endings and re-beginnings that happened for a lot of individuals that weekend, and that continue at an accelerated rate today. The changes are the changing of women's consciousness, the often difficult transition from patriarchy to a Goddess New Age. They are changes in individual psychologies that are basic and ongoing.

For most women, these changes consist in the emotional needs to let go of old "stuff" that is no longer working in one's life. Whether this be an obsolete idea pattern, a relationship that's no longer viable, a job or career or place of residence that is no longer positive, or a need for healing that will no longer wait, the changes have been profound. Sudden breaks with the old happened for many women during the actual Harmonic Convergence weekend or the week after it. Continuing and progressing changes are still happening, and increasing. The changes in individual lives have been difficult, fragmenting and traumatic, and the results have been growth. Through them women are finding their strengths and becoming more of who they are. They are discovering their consequence, learning their power-within, and manifesting surer and more whole lives in the process. The more psychically/spiritually sensitive the woman, the deeper the changes in her Be-ing.

Each Full Moon since the Harmonic Convergence has added to and speeded up the energy of August 17, 1987. For women on the paths of growth, the pressure to change has been tremendous. Some have left the planet, completed their work here and died, leaving others grieving. Most have died metaphorically in their old selves and been reborn as someone new, someone stronger and much changed. When old patterns end, a woman can resist the changes or let the old ways go. There is a need for something to replace them; if the old didn't work, what will? The easiest path is often the hardest, that of accepting and trusting and waiting for clarity. In letting go and in accepting something else more positive, there is growth and often pain.

Change begins in the individual, spreads to the group, and group change effects the collective unconscious, which potentially changes the world. On November 10, 1989, the Berlin Wall fell, ending the separation of Eastern Europe from the rest of the world and instituting major changes in the communist regime. East Germany, Poland, Czechoslovakia, Bulgaria and Rumania rapidly toppled their communist governments. In the spring of 1990, Lithuania, Latvia and Estonia declared their independence from Russia. In February, 1990, South African freedom fighter and leader Nelson Mandela was released from prison after thirty years. Change that begins in the individual increases power to change the world. Major healings have begun on planet Gaea.

By this process of change, of releasing the old and learning how to activate the new, the New Age begins. We are in the transition time now. Whether the transition to a New Age will be violent or peaceful largely depends on women, for women have the awareness to visualize and then actively create a positive new world. This is not to say that women are responsible for patriarchal violence, but that women are the only hope of changing it. As each of us changes within, the Goddess New Age moves closer to manifesting. As each of us learns the awareness to "harm none" and to create a peaceful world, the energy of the patriarchy begins to dissipate for the first time in 5,000 years. More and more women are learning that they have consequence, and are learning to use it for the good of all.

In this time of transition, of earth changes, women are relearning the skills that were lost under patriarchal suppression. More and more women are discovering the Goddess and all of the life-affirming values that She represents. Women are discovering their psychic abilities, long suppressed by male religions working to keep women disempowered. Women are learning about healing, the nonmedical system ways of caring for our bodies without drugs or technological intervention. We are learning ritual, meditation and visualization, ways of creating the world as we choose it, and are putting the skills to good use. We are learning to honor the Goddess, the earth, and each other, for the good of all. All change starts from within and women are changing the earth, with the approval and help of Gaea, who is changing with us.

As women change profoundly from within, they become midwives to rebirth the planet. She too is changing, is giving birth to a New Age, a new world. What that world becomes is something for women to create, to do ritual and to do political activism to achieve. The earth's changing is at a starting place, and she needs women's help. Pollution by chlorofluorocarbons (CFCs) is destroying the ozone layer that surrounds the planet, causing severe weather and a startling increase in skin cancer. Toxic wastes are poisoning the water supply. Industrial emissions that cause acid rain are deforesting the earth and polluting the air, adding to the ozone depletion and lung cancer. The soil is worn out from erosion and chemicalization, and with floods, storms and more frequent droughts, food shortages are happening worldwide. The earth's supply of petroleum is running out. South American and Asian rainforests are being cut down to graze beef for fast-food hamburgers. Hospital waste is washing ashore on public beaches, and dolphins are dying of an AIDS-like dis-ease.

There is much for women to do to create the Goddess New Age and to midwife the earth through her changes. Begin by becoming aware of the issues, the political, social and environmental problems and how they can be solved. Learn enough to be part of the solution on the physical level — learn about recycling and participate. Learn what corporations are the worst offenders and boycott their products. The worst offenders of the environment are usually the worst offenders in sexism and racism, too. Learn the simple ways that help the environment and implement them. Learn how life choices have an impact for good or harm on the environment, and learn to live more simply. Continue by working for civil rights — women's rights, lesbian and gay rights, anti-apartheid, anti-racism, reproductive freedom, disability rights — there are lots of issues to work on. March, do political actions, and keep informed of current events.

Next do the work from an inner level. Learn the power of visualization and manifesting and use it to work change in the world. The healing energy of many women went into the opening of the Berlin Wall. Learn about Goddess-within and women's consequence and apply them to personal and global life. Make working with these positive energies a part of daily living. Meditate nightly on peace in the world, send energy to Gaea for her healing; choose specific places or issues or use the energy in general ways, directing it to where it's needed most. Learn Goddess skills, concentrating on the skills that are most attractive to the individual. Some women will choose tarot, while others are healers; some are ritualists and others are artists. Any Goddess skill is a stimulus for psychic development, an important sixth sense for the women's New Age.

Most importantly, work on healing yourself. Every woman that has grown up in the patriarchy has been damaged by it. Our mothers have been damaged and have passed their pain onto us. No one is free of patriarchal patterns that need changing, whether the issue is to quit smoking or to come to terms with incest, to come out as a lesbian or to end an abusive marriage. Women go through their emotional earth changes and are being reborn in wholeness. They have met the pain, confronted the fears, left the wrongs, changed the old ways to become new. They have met their trials by cleansing fire to become women of the New Age, women whose new knowledge is the foundation of a utopian, matriarchal world. Once healed in themselves, these are the women who teach others, help others, and the numbers grow to become communities, and the communities band together to change nations, and the nations learn to cooperate to change the world.

Prophecies from channelers and New Age teachers of several years ago predicted great devastation in the world around the turn of this century, including a nuclear war. In the last three years, however, women have opened up their consciousness and awareness and rapidly made changes in themselves and in the world. As the awareness grows, the predictions soften. Most psychics now say that a nuclear war will not happen; our visualizing of peace has changed the world enough to prevent it. The advent of Mikael Gorbachev in Russia is a living hope, with wishes for American leadership to grow likewise. As women change they change the course of planetary karma, planetary herstory, to bring in a Goddess New Age as gently as possible.

This book is a part of that planetary change, a source of awareness for women creating a Goddess/matriarchal New Age and going through changes in themselves. In it concepts that Women's Spirituality may not have seen before are presented to focus on women's role in creating a new world. Much of the information is filtered through the popular New Age, reinterpreted. Much of the information changes radically when women's Goddess consciousness is brought to bear upon it.

Dreaming the Past, Dreaming the Future: A Herstory of the Earth is also a herstory of women's past. It's not a herstory to be found in patriarchal history books — women don't appear there much — but it's ours and needs examining. While some of the information is speculative, there is more evidence of its truth than of its fantasy. Moving from creation through the past and present and into the future, this book may be a herstory for the women of the Goddess New Age to come. The material on the future is necessarily speculative as well, and discusses some of our options for the world we now create.

Whether the earth changes we have entered will be peaceful or violent, a best-case or worst-case scenario or something in between, no one can say until it's over. I can only present the ideas, and sometimes my own opinions, and wait for the New Age to be. There is a Chinese curse that says, "May you live in interesting times." All of us on planet Gaea have chosen to incarnate this lifetime in highly interesting times.

Notes

1. Elinor W. Gadon, *The Once and Future Goddess*. San Francisco: (Harper and Row Publishers, 1989), p. 229.
2. José Argüelles, *The Mayan Factor: Path Beyond Technology*. (Santa Fe: Bear and Co., 1987), p. 159.

The Past

Women's Creation
The Beginning

The universe exists as sleeping darkness, unknowable, unknown, wholly immersed in deep sleep. Does she dream in sleep or only when she wakes? We know not. She sleeps. And then in her sleep the divine self appears with passionate creative power. She stirs, dispelling darkness. She who is subtle and full of desire, imperceptible and everywhere, now and eternal, who contains all created beings, wakes — and then the world stirs.

Monica Sjöö[1]

Creation has been female for the greater part of herstory, only becoming male in name in recent patriarchal times. The Goddess of a thousand different cultures and countries arose to consciousness and gave birth. The birth was first to herself, as all birth begins with awareness/idea/thought-form — the 'word' of the later Bible. After birthing her own Be-ing, the Goddess gave birth to the universe and earth, and then to all living things. In some cultures she birthed directly from her womb and vagina, as women have birthed ever since. In some she birthed a cosmic egg, from which all life emerged. In some she birthed the world snake who crushed the egg, spilling its contents into creation. In the earliest cultures, the Goddess is parthenogenetic, birthing without mating, and in later ones she creates her mate — sometimes male and sometimes the female image of herself — becoming pregnant thereby. Closest to recent times, some cultures changed the female egg to mean male generative power (notably in Egypt and India), but no woman yet has ever seen a male-laid egg! *"God was female for at least the first 200,000 years of human life on earth,"* [2] and she birthed all life from herself, as women do eternally.

Here are some of the stories of Goddess creation, the original genesis, beginning with Ilmatar of Finland. Ilmatar floated in the ocean for centuries, carried to the four directions by the ocean waves. A teal seeking rest built her nest on the Goddess' raised knee. Unable to hold her knee bent forever, Ilmatar moved, spilled the nest, and the eggs fell:

causing them to shatter into fragments.
From the lower shells the earth took form,
from the upper shells came the arch of heaven,
from the yolk came the lustrous sun,
from the white part came the moon,
and from all that was speckled in the eggs —
the stars came forth.
Still Ilmatar floated on the waters,
now peaceful and serene.
For ten more years She floated,
until the day when She raised her head
from beneath the waters —
and thus began Creation.[3]

Spider woman, creation mother of the Pueblo, Hopi and Navaho peoples, sang into Be-ing her two daughters, Ut Set who became mother of the Pueblo people, and Nau Ut Set, who became mother of all others:

Following the directions of Spider Woman, these two daughters formed the sun from a white shell, red rock, turquoise and pearly abalone shell. When it was ready, they carried the sun to the top of the highest mountain and dropped it into the sky so that it would give light. But when they saw that it was still dark at night, they then formed the moon, putting together pieces of dark black stone, yellow stone, red rock and turquoise. Still things were not quite right....It was for this reason that with the help of Spider Woman, Ut Set and Nau Ut Set created the Star People, giving them sparkling clear crystal for eyes so that there would never be complete darkness again.[4]

Another ancient creation legend is the Pelasgian (ancient Greek) story of Euronyme:

In the beginning, Euronyme, the Goddess of All Things, rose naked from Chaos, but found nothing substantial for her feet to rest upon, and therefore divided the sea from the sky, dancing lonely upon its waves. She danced towards the south, and the wind set in motion behind her seemed as something new and apart with which to begin the work of creation. Wheeling about, She caught hold of this north wind, rubbed it between her hands and behold! the great serpent Ophion. Euronyme danced to warm herself, wildly and more wildly, until Ophion, grown lustful, coiled about those divine limbs and was moved to couple with her...so She was with child.[5]

This version of herstory came late enough that the male is included. The earliest creation stories were female-only.

Yemayá is the Goddess of women and birth from Yoruba West Africa (Nigeria), and one of the Orishas of the Santeria religion of Spanish North and South America. She is the mother of all the gods and Goddesses, and the flowing of her birth waters created all the oceans and waters of the world. In this also later legend, Yemayá creates the sun, moon and stars. Other Yoruba stories

16

name the androgynous Oludumare (pure energy or Aché), or Obatala (who has both male and female aspects) as creators of the universe and earth.

> Yemayá gave birth to the sun, the moon and the stars after a brief dalliance with Olofi. As a gift for the magnificent children she bore him, Olofi gave her Ochumare, the rainbow, to wear as her crown. That is why the rainbow only appears when Yemayá, as rain, has fallen upon the earth, and her child the sun shines through the clouds.[6]

Virtually every culture whose creation stories are recorded has a similar explanation of the beginning of life. The Goddess created the earth, moon and stars, and created or birthed all life from her body. The oldest stories are female-only, with the Goddess the only creator. Later ones begin to include the male, first as a serpent and then in human form, her son that becomes her lover. In still more recent times, the son is her consort and equal, and then he begins to supercede her. The latest creation stories of all are those of male generation, with the female unmentioned. Many early Goddess stories were rewritten, with a female divinity suddenly made male, or with such interesting effects as the egg becoming a male symbol.

In the genesis of people on the earth, the stories are also ones of birth from the womb of the Great Mother. In the Mesopotamian version of creation, Ninhursag makes life:

> While the Birth Goddess is present,
> Let the Birth Goddess fashion offspring,
> While the Mother of the Gods is present,
> Let the Birth Goddess fashion a (human)...[7]

Puana of the Yaruros people is the Venezuelan snake Goddess, the snake a worldwide symbol for the female, birth-giving Great Mother. With the advent of patriarchy, the snake began to get bad press, as did any female or Goddess symbol. Later yet, it became a phallic symbol.

> At first there was nothing. Then Puana the Snake, who came first, created the world and everything in it....Kuma was the first person to people the land....Everything sprang from Kuma, and everything that the Yaruros do was established by her.[8]

Spider Woman is credited by the Pueblo and Navaho peoples as the first creator of people; another name for her is Thought-Woman. The first humans were born with a thread reaching from the crowns of their heads to the Goddess' vagina. When they asked her what it was, she told them that as long as they kept the thread clear and the crowns of their heads (the *kopavi* or crown chakra) open, they would always be part of her.

> Spider Woman gathered earth, this time of four colors, yellow, red, white, and
> black; mixed with *túchvala,* the liquid of her mouth; molded them; and covered
> them with her white substance cape which was the creative wisdom itself....She
> sang over them the Creation Song, and when she uncovered them, these forms were
> human beings in the image of Sótuknang. Then she created four other beings in her
> own form. These were *wúti,* female partners, for the first four male beings.[9]

Some of the continuity and breathtakingly beautiful cosmology of the Pueblo
people is in the stories of Spider Woman. There is more than birth here; the
connection of each Be-ing with Goddess is a part of the plan of the universe.

Creation is indeed the beginning of a vast plan, both the creation of the
cosmos and the creation of people. We were not put on this planet alone, to fend
for ourselves in struggle and confusion, though this is what it may seem in the
patriarchy's limited worldview. Women of today have very little connection
with a greater plan, little understanding that there is more than what exists in
physical bodies on a physical planet, for a limited physical lifespan. Perhaps the
greatest harm that patriarchy has done has been to separate the human soul from
the body, and to hide the meanings of creation and life from those who are
incarnated on earth.

Reincarnation is a piece that is missing from the cosmic puzzle, a central
concept of any matriarchal religion past or present. Reincarnation theory was
once part of both Christianity and Judaism but was dropped from both. The
Judaic reason for stopping to teach it was twofold: it was a tenet of the Goddess
religion, and it was deemed an idea too esoteric for the uneducated masses and
was kept for a long time among scholars only. Christianity dropped reincarnation from its doctrine in about the third century, at the same time that much other
early teaching was also dropped, including the Essene and Gnostic gospels that
promoted the equality of women. A knowledge of rebirth made people harder
for the church to control, it gave them consequence and autonomy. When
viewed with a background of reincarnation theory, the death and rebirth of
Christ takes on new meaning, and takes it closer to its pagan roots. In some
Middle Eastern Goddess cycles, the god who is the child of the Mother dies
yearly to become the life-giving harvest. He is reborn again at the new year. The
story is a later version of Demeter and Persephone, reflecting the addition of the
male as patriarchy took hold. The Egyptian cycle of Isis, Osiris and Horus was
Christianity's direct model.

The importance here is the concept of reincarnation itself, coupled with the
idea of women remaining connected to the Goddess who birthed them, and the
idea that life on earth is only a small part of existence. These were ideas that
went together in early Goddess cosmologies and were lost with the take-over of
the male religions. Reincarnation means that the soul is eternal, that death is
followed by rebirth into a new life. This is not to be confused with the Hindu
idea of transmigration of souls that says a human soul can reincarnate in a

nonhuman body. (A soul can do this by its own choice, but this is very rare.) Reincarnation means that the human soul repeats its existence in human bodies many times. Death is not an ending, but a continuing, a change of state from one thing to something else.

The soul reincarnates to continue its process of learning. It enters a physical body to experience living on the earthplane. When that lifetime ends, it returns to the Goddess for a period of evaluation, rest and understanding. Then it incarnates again, in a new body at a new time and place, for further lessons in being human. The accumulation of lifetimes becomes a body of knowledge; the soul does not repeat its lessons once learned, but moves on to new ones so the experiences continually change. In one life a woman may be African and live in a tribal setting as a farmer and leader; in another life she may reincarnate in China as a wife whose movements are restricted as she bears many children; in another life she may be American, and a doctor or engineer in modern times. She may reincarnate as male or female, and the roles she experiences change radically from one incarnation to the next. The purpose is always for learning what it is to be human and the end goal is a full understanding of both human nature and beyond-life cosmology, what in India is called "enlightenment." When that is attained, the soul is no longer required to reincarnate.

The central vehicle for connecting the learnings of one lifetime to another is the concept known as karma. In this idea that is so often misunderstood or only partly understood, the achievements and errors of past lives are brought from one life to the next as an accumulation of growth. Achievements are not lost and errors are places where more learning is needed, so that situations repeat until they are released by understanding. Karma is neither good nor bad, it is only experience. Channeler Mari Aleva defines karma as "choice and change." Each incarnation contains a series of opportunities for learning things unfinished from other lives. The goals of achievement are not in becoming rich but in becoming good, in understanding that all life is connected and behaving accordingly — "harm none" and act with consequence for the good of all.

The soul incarnates on planet Gaea for the purpose of doing good in the world. Since all life is a part of the Goddess, actions that "harm none" help everyone. The realization of connection, that all life is Goddess, and the serving of the life force while learning about oneself and one's place in the cosmos is the reason for incarnation and reincarnation. Things that hold the soul back from this realization of oneness are the situations that are repeated until they are resolved, until the soul learns the errors and changes her ways of relating. Pain learned on earth must be released from holding back the soul's progress — as in an incest survivor going through the difficult process of emotional healing. In the psychic idea that all time is one, some theorists believe that all of one's lives are happening at the same time, only *perceived* in a past, present, future timeframe. When out of the physical body, time does not exist for the soul.

The popular New Age idea of blaming someone who has experienced pain by telling her she must have caused pain for someone else in a past life is too simplistic. The woman who was raped was probably not a rapist in a past life. Yet her rapist may have been someone from a past incarnation who did wrong to her, and was not prevented from doing it again. The woman who was raped may have agreed in the between-life state (totally nonconscious on the earth level) to experience the rape for one or more reasons. Perhaps in healing the effects of the rape by therapy, she receives more wholeness than her life held before the tragedy. Perhaps by being the woman to prosecute the rapist, she does a service to many others by preventing him from doing it again. And if he escaped from punishment in the past incarnation, putting him in jail severs the karma between them. Sometimes even great pain can have its reasons for being, reasons that may be positive in the wider view.

The rewards of karma are also very real, although less recognized. Souls reincarnate in groups and a lover, friend or child can remain a companion and a comfort through many lifetimes. Things learned and enjoyed in one incarnation can carry through to other lives; look for them in skills learned seemingly effortlessly and mastered quickly. When a child exhibits talents well beyond her training or years, this type of karma is probable. Karma offers no more lessons in one lifetime than the soul is willing and able to experience, and there are always joyful lessons with the hard ones.

The learning, if not the actual experiences and lives, is meant to carry through and connect the lifetimes. After death the soul evaluates its growth for that life, compares and adds it to learnings from other lives, and chooses what it needs to learn next in coming incarnations. The period between lifetimes is a resting place, a going home, where all of the confusion of being in a body is healed, all the mental, emotional and spiritual experiences integrated. It's a place of joy, and a place of returning to a soul family, where others that one loves are waiting. When a soul reincarnates, it chooses the circumstances of that reincarnation with the help of guides and the soul's spirit family. It chooses the time and place, whether the incarnation will be male or female, rich or poor, what opportunities will be presented, and it chooses its parents. Each choice is based on what the soul can learn from the circumstance.

In the work of channeler Mari Aleva, when a child is born, the incarnation puts only one of many light-strands of her soul into physical form. The core soul is like a spiraled helix, a DNA rope of twined-together strands of light and energy. One multi-faceted strand becomes the life force of the child, another one becomes that child's spirit guide. Some of the other strands may incarnate while the child is on the earthplane; if any of them do, they are twin souls that she may or may not meet. The core soul, a part of which incarnates in a physical body, is the Goddess-within essence, the life force of the child that is born.

From Mari Aleva, channeling the energies of Teacher:

> Envision yourself as a light beam. Envision a cord, a rope if you will, entwined with many faceted strands. Each of these multi-faceted strands has a life in and of itself. The combination of many of these cords within your own centered light being is what we call your core self.
>
> Core self is your own inner being. It is a combination of all you are in this life and beyond. This consists not only of past lives, but other existences as well. Yes, we are speaking of other dimensions beyond the planetary earth plane level. These are all the energies entwined, involved and engraved within each of our soul selves' core.[10]

Soul and body are both made of pure energy (light, or the African Aché), and their coming together creates consciousness, the awareness and thought form that creates the self and the world.

We are not meant to come to the earthplane alone, nor to lose our connection with the universal plan. No soul incarnates without one or several guides. It is a tragedy of life in the patriarchy that most aware children — and children are born aware — are separated from their spirit guides, their "invisible playmates" or "guardian angels" at an early age. They are told that their "friends" are made up, that they don't exist, that they are lying. Christian baptism further complicates things by symbolically closing the infant's third eye with the cross in the christening ritual, closing her connection with Goddess and the world beyond. New Age, wiccan and Goddess women are rediscovering their guides, their soul's connection with itself and the Goddess, and are regaining their awareness of the cosmos and their place in a universal plan.

But how did souls originate and why did they enter bodies in the first place? In the Christian-oriented words of Ann Valentin and Virginia Essene in their channeled book, *Cosmic Revelation* (SEE Publications, 1987), "The soul is a narrow beam of energy created by God's first two children, the Gold and Silver Rays."[11] Translate this in Goddess terms to mean that the soul is a Be-ing of energy and thought form, created by Spider Woman's two daughters. (Or pick another creation story of your choice, the Goddess has many names.) The soul is an energy form, a light Be-ing, as is all that is created in the universe and the universe itself — all is energy. Each soul is the basic and primary source of the individual, and it has existed beyond the physical body for actually millions of years. The soul is a portion of Goddess.

The Gold and Silver Ray (the twin Goddesses Ut Set and Nau Ut Set; Valentin and Essene call them male and female or active and receptive) created helpers for themselves and made the planet earth. It was made with the intent that the lives upon it would be interdependent, but also created with variety and free will. Souls to initially populate the planet/experiment were chosen from volunteers for the project, about a million of them who agreed to remain here and care for the earth. Nearly 80% of the first million souls were chosen from either the Pleiades cluster or the constellation of Orion, with some from Sirius and a few from other places.[12] The earth is not, and has never been, the first or only planet to support intelligent life.

Initially, souls did not incarnate in bodies, but began to do so only as the earth evolved. In the beginning, they moved in and out of light-bodies, and may have borrowed the bodies of incarnated animals, as temporary vehicles to experience the earthplane directly. In Valentin and Essene's story, one of the creators' helpers turned away from the designated plan for the planet, and manifested negativity in the souls. The souls, by accepting it, forgot their purpose on the earth and became entrapped in the material, less and less able to come and go. Their light-bodies became dense physical ones, their soul power greatly diminished, and they lost connection with the world beyond the earth. They lost knowledge of universal purpose and the Goddess' plan.

To lessen the entrapment in bodies and to bring the souls back to awareness of the Goddess cosmos, reincarnation was established and each soul was given a spirit guide or guides. Early lifespans were much longer than they are today, and between incarnations, souls return to their true power and form. The many teachers that incarnate on the earth do so for the purpose of raising souls' consciousness while embodied. When the earth learns peace and connection to the universe again, incarnation will no longer be required. For now, this learning is an individual process. Because the earth does not live in peace, harming none and for the good of all, we are quarantined from the lifeforms of other planets at this time. Women, the creations of the Goddess/Silver Ray, have remained more in awareness and lead the earth's movement now to return it to life-affirming ways. Matriarchal awareness is Goddess/soul awareness. The energy of the earth changes is promoting this plan.

Ruth Montgomery, pioneer of the New Age and a channeler by automatic writing, has a creation story similar to that of Valentin and Essene. In her book *The World Before* (Fawcett Crest Books, 1976), she describes a similar process of the creation of the earth and of people. Before there was the universe, there was "only a void." I have changed "god" to "Goddess" in this passage:

> Not even chaos existed, they say, because, "There was no sound, no vacuum, no stillness, and no sleep. The time for awakening was a time of Deep. Yet the Force that we call God(dess) was always, for without God(dess) there is no thing, no motion, not even a nothingness."[13]

From the awakening of chaos — the word means unordered potential — came Goddess as force and idea/thought form and all things came into Be-ing:

> As ideas formed in the Mighty Being, they became deeds. From molecules and atoms thought-into-being by this Force came specks and particles that gradually melded together and began spinning in space, and as direction was given to these revolving masses, there gradually evolved planets which swung into predestined

orbits around suns of magnetic force.[14]

The passage echoes the Goddess creation stories from very early cultures. Remember Euronyme.

Life developed on the planet, changing from the simple and minute to the more complex. In the following quote, I have made the male pronouns female, changing "god" to "Goddess."

> So intricate and exciting became this system of growth that the force we call God(dess) desired companionship to share (her) joy, and in a mighty burst (she) cast off trillions of sparks from (her) exalted Being, each spark a soul.... [15]

This also sounds like the ancient Greek story of Euronyme. No new souls have been created since this beginning, according to Ruth Montgomery, and some souls have never incarnated into bodies.

The earth evolved into growth and bloomed with plants and animals. Souls, anxious to experience life on the earth directly, discovered they could move in and out of their light-forms and enter the bodies of the animals.

> Some of these curious souls experienced the thrill of eating berries, fruits and nuts for a time, and then withdrew to spirit form, leaving the animals unmolested. Others so greatly enjoyed the experience of procreating, eating, and sleeping that they became entrapped, and were unable to leave the...physical bodies. [16]

When the situation erupted into jealousies and competition, Goddess thought into form human bodies, giving them free will and minds able to choose between good and harm. Some of the now-humans, against warning, mated with those souls trapped in animal bodies and propagated. Montgomery describes this as the original fall from grace in the Garden of Eden, original sin. The half-human, half-animal Be-ings of ancient mythology were more than legends, they were real, and became subject to the first examples of racism later. The first entirely human Be-ings were called Amelius, and were androgynous, both male and female in one. These were who the Old Testament calls Adam and Eve.

> After human souls were separated into male and female so that they could produce their own kind, Goddess imposed Divine laws making it impossible for human beings to produce offspring as a result of cohabitation with any other species. Five races of humans simultaneously entered the earth....Amelius returned as Adam-Eve hoping to demonstrate that spirit could live in physical body without greed or envy...in perfect harmony. In this way the earth was peopled. These spirits in human body were able to communicate with each other by thought...ESP; and they were also able to free themselves of earthly forces in order to create giant objects and seemingly lift them from one place to another. [17]

More will be said about the earliest of civilizations in the next chapters, and about the half-human, half-animal Be-ings that remained. Throughout her account, Montgomery emphasizes the more-than equality of women in the scheme of things. "Women were as active as men in producing the wonders of an ancient age..."[18] Women in fact, were dominant in the establishment and ordering of civilization on earth.

From the work of Mari Aleva comes more on how souls initially entered bodies in the evolution of the earth. All of the material on this subject has been channeled, brought from the Akashic records, the soul records of human incarnation, through the minds of women now alive on the planet. The interpretations are theirs, and filtered through my own understanding and awareness for this book. The information is speculative; no one can prove or disprove it. I present it for women's education and deep thinking about where we come from. Listen to your inner self for what feels right, and apply to the material your own ability to manifest reality by thought. There are no right or wrong answers, there is only thought.

Mari is a conscious channel, relaying the energies of several entities. Teacher spoke through her earlier in this chapter, and speaks again with this poem on creation:

In the beginning there was the Dark of Womb,
Beginning of Thought,
Projected from the Womb of Life
And there became Life,
Pulsing, vibrant life energy
Birthed unto all in the form of Thought.

Thought Projection activates Life Force as we now know it.
That pinnacle of Attraction-ment
Is the Basis of all Life Forms as we realize them,
And the Thought itself is the
"Cosmic Glue" within us all.

Before beginning this book, I asked Mari Aleva for a channeling session and we taped it at Candlemas, 1990. Caridwyn Aleva transcribed the tape, and portions of the session appear throughout this book. The full transcript is in the Appendix.

Ruth Montgomery states that the first souls evolved into bodies on the early continents of Mu and Atlantis. These earliest civilizations were matriarchies, and are the subject of the next chapter. Mari Aleva, channeling a female entity named Helm, also describes souls' first entrance into bodies as having occurred on Atlantis, and she discusses cellular memory reaching back from now to that point. Helm speaks with an Eastern European accent, and is a delight to talk with. I am grateful to her and to Mari for their input in this book, and particularly for their ideas on early matriarchy. From Helm:

24

There are many times that people here smell a certain aura, aroma, and it smells very strangely familiar and yet it is new. It brings back memories, cellular memories. You see we believe what we teach is the body itself has memories beyond your own soul self body. What we are speaking of is the DNA from your parentage all the way down from your parents and their grandparents. You have all of these memories that go down what you call your bloodline.

There are also what you would call spiritual connections which are the essence of your core soul being before and after it combines with your body, and what combines it would be mental attitudes and emotional input. The spiritual part transcends way beyond what you call this earth. Much of the cellular content as well is beyond this earth, for it is our belief that *there have been people come to earth that have not evolved from the amoeba. They have been sent here in what you would call a spaceship.* There are many things connected with what you call spaceships having to do with inter-dimensional travels and bringing their vibrations in connection with the earth so that they can be here to live and to colonize.

Diane: People came to earth initially from other dimensions? Is that what you're saying?

Helm: There are people here, yes.

Diane: Was earth settled that way? Did humans come here intially that way?

Helm: Certain parts of the earth were settled this way. Some of this, what you call Atlantis, is before earth change here. Women in ancient civilizations, that Helm knows, has experience with.[19]

I have emphasized the most startling information, and yet as information unfolds, it becomes a familiar theme. The earth change mentioned by Helm ended in 10,300 BCE with the final sinking of Atlantis. These were the first souls manifested into bodies. How does a soul incarnate today? Again, from Mari Aleva:

One strand of your core reaches out to the earth vibrations in "this" universe and begins to vibrate to a similar beat as the drummer of earth's vibrations, and...here we are!

A body has manifested. Now remember that in the beginning as our core soul we assume the reality of pure spirit, and now something new has been added to the particular strand that has projected out to the earth vibration — a body — the vehicle by which we use and need to stay in the earth atmosphere...The body is what keeps this strand of our core here.

The bodies that we have manifested are also pure energy....As the body and spirit merge, a new element is created, and that is what we call the conscious aware mind.[20]

Psychically derived knowledge from more than one source suggests that people, or at least some of the people of earth, come from other dimensions or solar systems. Mari Aleva's channeled information states that some of earth's early inhabitants did not "evolve from the amoeba," but were brought here in spaceships. By translating literally the early written tests of ancient Sumer, a corroborating story comes forth. Biblical scholar Zecharia Sitchin tells this story, translated from ancient texts and modern commentary in his thought-provoking book, *The 12th Planet* (Avon Books, 1976).

The founders and teachers of Sumerian civilization, as their own records state, were the "Nefilim," a word that scholars generally translate to mean "giants" or "the children of the gods." The word, according to Sitchin, literally means "those who were cast down," as in "those who were cast down upon the planet," or "those who came down from heaven" — in spaceships. Other writing and pictorial records verify this interpretation, including a word in the language for "spaceship," "shem." The word "god" in Sumer was usually used in the plural, and applied to those Be-ings who arrived on earth and built an extensive mining and oil drilling operation here. They were male and female, technologically advanced, and apparently quite human. The gods of several cultures, including the Greek pantheon, may derive from these space-people.

Where did they come from? The Sumerian creation epic, the *Enuma Elish,* explains how the world came to be, and features the Goddess Tiamat. It's a tale of war, how Marduk defeated and dismembered the Goddess so that she split in two, one part becoming heaven and the other earth. Women have read the story as a metaphor for the takeover of the Goddess and matriarchy by male aggression. In Sitchin's interpretation, however, each of the characters in the Tiamat story is the Sumerian name for a planet, and there were twelve planets (including the sun and moon) in the Mesopotamian solar system. The extra ones were called Tiamat and Marduk. By this reading of the *Enuma Elish,* the tale that emerges as the creation of earth from one of the halves of the shattered planet Tiamat, is an actual cosmic happening put into metaphor.

Tiamat was the planet located in the solar system between Mars and Jupiter, where today there exists an asteroid belt (in Sumer called Heaven's Bracelet) of fragments that scientists believe was once a planet. At the time of the occurrence, the planets were not fully stabilized in their orbits. A new planet from the furthest reaches of the solar system, beyond Neptune, was formed and entered into the gravitational pull of the sun's system. In its comet-like passage, it disturbed the magnetic/gravitational pull of Tiamat, passing too close:

> An unstable solar system, made up of the sun and nine planets, was invaded by a large, comet-like planet from outer space. It first encountered Neptune; as it passed by Uranus, the giant Saturn, and Jupiter, its course was profoundly bent inward toward the solar system's center, and it brought forth seven satellites (moons). It was unalterably set on a collision course with Tiamat, the next planet in line. [21]

Tiamat and Marduk did not collide at first, but a satellite of Marduk smashed into Tiamat, "vanquishing her." Marduk continued on its orbit. On its next sweep through the center of the solar system, Marduk itself hit Tiamat, splitting her in two, and shattering one of the halves. The force threw the other half out of its orbit and Tiamat, with her moon Kingu still attached, became the earth. Says Sitchin:

> We are offered — for the first time — a coherent cosmogonic-scientific explanation of the celestial events that led to the disappearance of the "missing planet" and the resultant creation of the asteroid belt (plus the comets) and of Earth. After several of his satellites and his electric bolts split Tiamat in two, another satellite of Marduk shunted her upper half to a new orbit as our planet Earth; then Marduk, on his second orbit, smashed the lower half to pieces and stretched them in a great celestial band.[22]

Tiamat reestablished her orbit, axis, spin and gravitational pull. Earth's continents are all on one side of the planet, and the other side is a deep cavity (the Pacific Ocean bed). Marduk left the solar system center again; Sitchin gives it a 3,600 year, very wide orbit. Both planets evolved life, or were settled from elsewhere — Marduk may have "seeded" Tiamat/Earth with life in their encounter — and Marduk became technologically advanced. In later times, its orbit passing near again to the center of the sun's system, ships from Marduk landed in Sumer.

The Sumerians describe a full history of their land when it was inhabited only by the Nefilim. They state in several writings that the human race was created by these "gods" to do their work for them. The word "Adam" from "Adama" (earth) suggests that this creation of the gods was an "earthling" or "earthborn." Science says that *Homo sapiens* evolved, but has no evidence of the steps. Anthropologically, she "just appeared." If she were created by the genetic engineering of a technologically advanced people, that sudden appearance is explained. As incredible as it sounds, Sitchin has the written and linguistic proof from the ancient Sumerian texts for his theory, and describes the process by which *Homo sapiens* was created.

> The Nefilim did not "create" man [*sic*] out of nothing; rather they took an existing creature and manipulated it, to "bind upon it" the "image of the gods"....
>
> For, some time circa 300,000 years ago, the Nefilim took ape-man (*Homo erectus*) and implanted on him their own image and likeness.[23]

There were mistakes along the way in this process of merging two lifeforms, and ancient historian Berossus describes some of the half-human, half-animal mixtures that Ruth Montgomery wrote of. They are the creatures of mythology, human heads with animal bodies, creatures with multiple heads, or wings, or hermaphrodites (having both male and female organs). Apparently, Nefilim scientists first tried to make their workers non- or semi-human. There were also six all-human attempts that were nonviable. In their quest for workers, they tried every aspect of their advanced genetic technology, until finally succeeding with Adama, a hybrid of *Homo erectus* and the "gods." Adama was born from the womb of Ninhursag, and thirteen other surrogate mother Birth Goddesses were enlisted in the project. (See the passage about Ninhursag creating people at the beginning of this chapter.) Once created, s/he proved to be genetically compatible with her creators. Much later the Nefilim left earth, and their creations took their independence.

There are echoes of this story in the biblical Genesis, if literal translation is applied. There are also echoes of some of the Goddess creation legends, which were masculinized later. Though the actual origins of people cannot be proven at this time, the version known almost exclusively in the West is not the only version and almost certainly is not the whole story. It was first in the Middle East that creation became male, or earlier male and female, where all other cultures initially saw creation of the universe, earth and people as female-only. Until the advent of the genetic engineering of the Nefilim, birth occurred only from the womb, and even the Sumerian Adama emerged from there eventually. As patriarchy grew, reference to the Goddess birthing the universe was changed to a male god's creation.

In accord with Zecharia Sitchin's theory of space visitors, Merlin Stone in *Ancient Mirrors of Womanhood,*[24] describes the arrival of the Goddess Ashtart as a fiery star landing on Lake Aphaca. The name is connected with Ishtar, possibly another name for the Sumerian Inanna, and another Great Mother of the Middle East. Inanna may have been one of the Nefilim from Marduk. Ishtar is said to have come to earth from Venus. The Goddess Aphrodite, usually associated with Venus, was a Canaanite Goddess originally, and Canaan is one of the later cultures to emerge from Sumer. The Dogon people of North Africa describe their origins as from the solar system of Sirius, the Dog Star, and connect her with the Egyptian Isis or Sothis. Ancient Mayan art depicts astronauts. The Nefilim, or their space-traveling sisters from other planets, may have brought the Great Goddess to earth.

All creation begins with thought, whether it be creation of a universe or of people, or of an individual soul entering a body to incarnate, or of any creativity on earth. Thought is the creative force of the Goddess, of the universe, and of women. Spider Woman — who carries the other name of Thought Woman — sang her idea of creation into Be-ing, birthing her two daughters who formed the earth. She created people in much the same way, and in her image. The female thought to create makes universes happen, and the psychics say that *all* is energy and thought. As women, we daily think our worlds into Be-ing, our lives into Be-ing. If thought created the universe, and thought manifests our souls into bodies, what else can it do? What *can't* it do, if we learn to use it?

Women were the creators of the universe, of the moon, sun and stars, of the earth and of all the life on the planet. Without our wombs, nothing could exist. With our ideas, anything can be made manifest. We are the creators, the thinkers, whose women's values can change the planet to a place of peace, to a garden of delights. We are the women who will one day again reach the stars; we are the daughters who once came from them. In legends from Greece and China, women became stars — or was that a confusion of the idea that some women may have come *from* them?

The material in this chapter is only a beginning of the herstory we have never heard before. Time has distorted our real beginnings, and patriarchy has rewritten them to serve itself. The creation of this chapter cannot be proven, but neither can the creation stories that have been written for us. We are daughters of thought and of the stars, and need to remember that, relearn that, to take our place in the Goddess New Age. There is another version of the idea of a male-created universe, that a male god made a male person from clay and breathed life into him by touching him. When women are giving birth all around us that story pales to insignificance. The universe and all life came from female Beings, from Goddesses' wombs. This is only one of the things that has been taken away from us, but we reclaim it. We are Goddess.

Notes

1. Monica Sjöö and Barbara Mor, *The Great Cosmic Mother: Rediscovering the Religion of the Earth* (San Francisco: Harper and Row Publishers, 1987), p. 55.
2. *Ibid.,* p. 48.
3. Merlin Stone, *Ancient Mirrors of Womanhood: A Treasury of Goddess and Heroine Lore from Around the World* (Boston: Beacon Press, 1979), p. 340.
4. *Ibid.,* p. 289.
5. Monica Sjöö, *The Great Cosmic Mother,* p. 57.
6. Migene Gonzalez-Wippler, *Tales of the Orishas* (New York: Original Publications, 1985), p. 26.
7. Translated by Zecharia Sitchin, in *The Twelfth Planet* (New York: Avon Books, 1976), p. 350.
8. Monica Sjöö, *The Great Cosmic Mother,* p. 58.
9. Frank Waters, *Book of the Hopi* (New York: Ballantine Books, 1963), p. 6.
10. Mari Aleva, "Your Core Soul Connection," in *Voices From Beyond,* (Westland, MI: Moonwind Publications, 1991). p. 41.
11. Ann Valentin and Virginia Essene, *Cosmic Revelation* (Santa Clara, CA: SEE Publications, 1987), p. 45.
12. *Ibid.,* pp. 48-49.
13. Ruth Montgomery, *The World Before* (New York: Fawcett Crest Books, 1976), p. 17.
14. *Ibid.,* p. 18.
15. *Ibid.*
16. *Ibid.,* p. 21.
17. *Ibid.,* pp. 26-27.
18. *Ibid.,* p. 24.
19. Mari Aleva, *Earth Changes with Mari Channeling,* pp. 3-4. See Appendix.
20. Mari Aleva, "Core Soul, November 29, 1989," from her book in progress, *Evolution of the Soul.*

21. Zecharia Sitchin, *The 12th Planet*, p. 224.
22. *Ibid.*, p. 227.
23. *Ibid.*, p. 341.
24. Merlin Stone, *Ancient Mirrors of Womanhood*, p. 103.

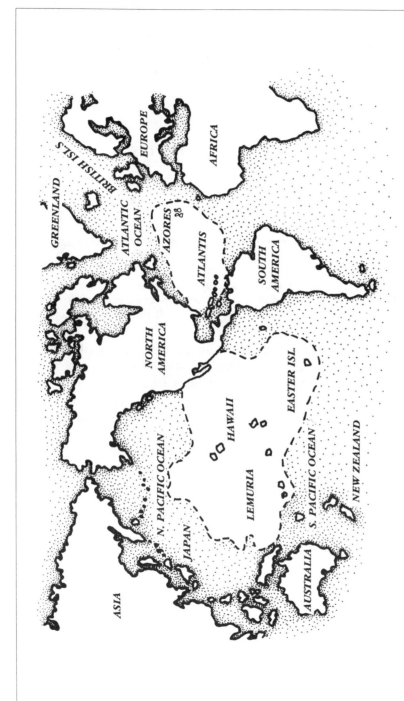

This map is derived from information in Earlyne Chaney's book, *Revelations of Things to Come*. I believe Atlantis to have extended farther west, connecting to, or very near to, Central America.

The First Matriarchies
Mu and Atlantis

The soul of the human is made of star matter, Pleiadean matter, Sirian matter, Uranean matter. The soul of the human is merely a collection of energy. The collection of energy of the soul, the being, is the same matter as the collection of energy of all energies — stars, galaxies, and supernovas. Know this fact in such a way that you have power to merge with this energy, and then you can know matter. . .

Barbara Hand Clow[1]

In the larger herstory of the world beyond Sumer, the first souls that entered bodies lived in Atlantis, or rather in its pre-civilization of Mu. What the people themselves called this continent in the Pacific is unknown, but it is today referred to as Mu, Lemuria or Pacifica. Remnants of it still exist in living Pacific islands, in huge monolithic stone statues and in cities under the sea in dozens of places all around the Pacific Basin. To give a time frame for this herstory, psychic Edgar Cayce states that people began incarnating on the earthplane ten and a half million years ago (10,500,000), and that most of Mu sank beneath the Pacific in the earth changes that expanded Atlantis to a continent, around 50,000 BCE.[2] The people who lived in Mu were also the people of Atlantis, their civilization becoming the Atlantean culture after Mu was destroyed. Mu was geographically connected to Atlantis through what is now Central America.

The herstory of Atlantis and Mu is given somewhat differently in this book than in traditional sources, with emphasis on the connection between the two cultures. According to Mari Aleva and other psychics, the first people came to earth from other dimensions or planets, and some of them arrived in Mu on spaceships. The Pleiades is the place most often mentioned. These were also the first people to incarnate in bodies as humans. All of the five races were created at once and they lived in the same land, initially as separate tribes or families. The culture was matriarchal and spirituality-centered in later Mu, and the people lived underground, first in caves and then predominantly in underground dwellings. Enough large and hostile animals existed to make this necessary; this was the day of the dinosaur and mastadon. Remnant examples of this early housing remain in the Hopi kiva and the gorgeously decorated rooms and temples of Çatal Hüyük. Do not mistake these people for primitive. They were the people who communicated by thought and ESP and who created and moved the huge monoliths of Easter Island. Their stone block sculptures, roads and monuments rest on the ocean floor all around the Pacific basin. They were newly embodied and still psychically connected to the Goddess universe.

The land of Mu was located in the Pacific, a continent that included what are now Australia, New Zealand, the Philippines, Oceania and western North America. These are all centers of early Goddess culture, and all are sources of Goddess creation stories. Col. James Churchward, who found records of Mu in a temple in India, learned to translate them and discovered a creation story that involved the cosmic egg, and a world symbol for Mother which also meant Mu. [3] Unfortunately all of the books available on Mu and all but Ruth Montgomery's *The World Before* on Atlantis are by men, and feature male interpretations. Here is a portion of the creation story, which was written as a series of seven commands, and which Churchward compares with Genesis:

> The fifth command was:
> Let life come forth in the waters. And the shafts of the sun met the shafts of the earth in the mud of the waters and there formed cosmic eggs...out of particles in the mud. Out of these cosmic eggs came forth life as commanded.

> The sixth command was:
> Let life come forth upon the land. And the shafts of the sun met the shafts of the earth in the dust of the land, and out of it formed cosmic eggs; and from these cosmic eggs life came forth upon the earth as was commanded....Then Narayana, the Seven-headed Intellect, the Creator of all things throughout the universe, created (wo)man and placed within (her) body a living, imperishable spirit, and (she) became *like* Narayana in intellectual power. Then was creation complete. [4]

The symbol for the creator Narayana was an egg with a sea-serpent inside it, both familiar Goddess symbols. I have changed Churchward's male pronouns to female in the last paragraph — it seems appropriate.

Churchward's material was translated from clay tablets and is a concrete written record, but his interpretation is highly patriarchal and he seems unable to separate Atlantis from Mu. Another major source for information is the Lemurian Fellowship, whose channeled lessons are quoted by David Hatcher Childress. I was not able to locate this original material, so I quote it second hand. Enough information on the beginnings, structure and government of Lemuria are clearly and logically presented by this source to make it pertinent. Lemuria is a name coined by later scientists, though many psychics use it. The Fellowship says the continent's real name was Mu or Mukalia:

> The Lemurian, or Mukulian civilization...lasted in the form of an empire for 52,000 years and reached heights so great that our present civilization can barely be

considered a civilization when compared to it. Government, religion, and science achieved such perfection as to be far beyond our present comprehension.[5]

Childress explored the Pacific basin looking for evidence of Mu, and found it extensive. Members of Undersea ruins exist showing similar monolithic architecture and evidence of highly developed cultures.

The Lemurian Fellowship dates Mu as younger than Edgar Cayce established it, stating that it rose about 78,000 years ago and was destroyed in 26,000 BCE. Based on several sources written and channeled, the information we have on Atlantis, and on geologic history, I believe the older dating to be more correct. The date of 26,000 BCE is usually given as the second of the three earth changes affecting Atlantis. Some portions of Mu may have remained until that time, but most sources place the first earth changes that destroyed Mu as the beginning of the rise of Atlantis around 50,000 BCE.

Mu was destroyed by a pole shift — a shifting of the axis of the earth — resulting in earthquakes that caused the continent to sink beneath the sea. Edgar Cayce says the shift was a technological accident that started with a world conference on methods to destroy the dinosaurs about 52,000 BCE. Explosives used on animal nesting places blew open gas pockets extensive enough to destabilize the earth. The planet's axis shifted, moving the poles to their present locations and initiating the last ice age, along with earthquakes and volcanic activity. The dinosaurs perished; the continent of Mu began to break up, with refugees migrating to Atlantis and South America.[6] Geologists state that the earth shifts magnetic poles, but not its axis in a pole shift and thus question the gas pocket theory.

When Mu sank, the world's oceans lowered because of the volume of water that rushed to fill the void where the land mass had been. (A geologist says that an ice age would be more likely to lower water levels than this explanation.) But whatever caused the lowering, what had been small islands in the Atlantic became the tops of mountains on a continent that rose above water. These islands were the Poseid Archipelago, and the continent, originally called Poseid, became Atlantis.[7] This contradicts Edgar Cayce, who says that the pole shift of 50,000 BCE began the breaking up of Atlantis instead of the breaking up of Mu. A portion of Mu's population had already founded a colony on Atlantis for political reasons. Now more survivors reached Atlantis and were the foundation for that country's government and culture. Those Mukulians who reached South America founded the Inca culture, and the North American Hopi people — with their teaching of the four worlds — may be survivors of Mu.

The continent of Mu contained twelve fertile valleys that were essentially isolated from each other, and over a period of half a million years the races of the world developed as tribes in these valleys. A group of female Elders acted as liaisons between the tribes, choosing promising candidates from all of them to train, including two with the names of Rhu and Hut. Using these two peacemakers, the Elders forged an alliance among the tribes for the settlement of the continent's great plains area, a very fertile land much contested.

The tribes agreed to cooperate and built the capital city of Hamakulia at the mouth of the valley. The Elders established a training program in cooperation and culture for those who would settle in the valley, and this became the beginnings of citizenship training. This training took seven years and the aspirants lived at the training schools; children were eligible for citizenship training at twenty-one years of age, so the graduated citizens were at least twenty-eight. To admit only those sincerely wanting citizenship, a student had to give the school half of her possessions in payment, as the school fed and clothed her for the seven years. Upon graduation, however, the student's herds were returned to her, upgraded by the Elders' breeding knowledge, and the new citizen gained more than she sacrificed.

Skills taught in the citizenship schools were the women's work skills of creating civilization — dairying, cheese and buttermaking, the use of milk in cooking. They learned the agriculture of vegetables and grains, hide tanning, wool processing by carding, spinning and weaving. The culture was very much centered on childrearing and this also was taught. Toolmaking, pottery and building were taught and small industries were developed.[8] To become a citizen of this fertile valley required the learning of civilization, cooperation and peace. The members of all tribes were eligible for the training.

Children were educated to the age of twenty-one, and then were eligible for citizenship training, which allowed them to take residence in the valley. Public office terms lasted for life and required another seven years training beyond citizenship. Ten laws were the basis of government in Mu:

1. No one shall profit at the expense of another.

2. No one singly, nor the commonwealth collectively, may take anything away from another by force.

3. All natural resources shall remain the property of the state or commonwealth, and may not be claimed as a personal possession....

4. Every citizen and every child thereof shall be entitled to and receive equal education, equal opportunity for the expression of ability and equal standing before the laws of the land.

5. All advancement in position shall be based on merit and the performance of service alone.

6. No individual shall be entitled to retain as a personal possession anything for which s/he has not personally compensated equal value.

7. No individual shall have the right to operate in the environment or personal affairs of another unless asked to do so by the person.

8. No one may intentially kill or injure another person, except in the defense of life or state.

9. The sanctity of the home shall be kept inviolate, and no woman may be taken in marriage without her consent.

10. In all matters affecting the common good, and when no violation of Natural law is implied or involved, the opinion of the majority shall rule, subject only to the consent of the Elders whose decision shall be final.[9]

Those who showed the greatest aptitude were trained by the Elders in spirituality, and it was this priestess-class of highly trained Be-ings that designed and supervised the running of society. They were adepts or "saints" who exemplified the goals of the Mu culture, the healers, civilizers and peacekeepers. Says the Lemurian Fellowship:

> They were loving, knowledgeable, kind and sympathetic human beings dedicated to helping others journeying along the same path. [10]

Channeler Mari Aleva talks of women in the ancient civilizations. Helm is speaking through her:

> The women were not only the nurturers, but the providers as well. They were the gatherers and teachers, for when the children were born to the women, the women would be the ones to teach and to bring up the family as the number one priority of the family. And then they say, well, who will take care of the children? You have extended families, sisters. You see, the men would do what their bodies do, the strong things, to build. In the old days here on earth you call that slaves. Only we are not saying men were slaves per se. But they would build. And there was also when there were invasions, the men would stand as guard. There are many ancient traditions who believed in the spiritual realities and they would train to protect against this. [11]

Through Goddess archeology, the work of such women as Marija Gimbutas and Merlin Stone, women are becoming aware of early matriarchal cultures. These have been excavated on every continent and display the tools and arts of highly advanced, ancient civilizations. They were peace-loving cultures, cities built without walls or defense and with no weaponry ever found in their artifacts. They describe an era of peace and well-being, an age of women and women's values and skills, a golden age to model the New Age upon. All of these cultures celebrated the Goddess and revered the life-giving values of women; they were life cultures very different from the death cultures of patriarchy. Was the early Goddess matriarchy that women seek to emulate today the culture of Mu and early Atlantis? Did it originate on earth at all? Helm has stated more than once, "People descended from other place, not evolve from here."

But Mu was not totally peaceful, either, despite the work of the Elders to make it so. The great plains were settled and civilized only after centuries of fighting over them. And even after the Elders established citizenship training there still were factions:

> The destruction of the Empire came with the allowance of non-citizens into the Empire. While they could not vote, they enjoyed all the material comforts of the citizenry. However, because they could not balance themselves as citizens did in their education, non-citizens tended to develop into either extremely practical or extremely spiritual personalities. The Lemurian Fellowship calls these two factions the "Katholis" and the "Pfrees." The "Katholis" prized their spirituality, while the "Pfrees" prized practicality.[12]

As the conflict became more disturbing to the daily life of Mu, large groups of mostly Pfrees either chose or were asked to emigrate. One tribal group settled on Poseid Island, which became the major Pfree colony. Others settled in China, the Gobi Desert, or South America and developed the cultures there. A Katholis group settled in India, becoming the Rama Empire and subject of the war between India and Atlantis that is described so graphically in both the *Ramayana* and *Mahabarata*.

When the land mass of Mu was broken up by earth changes around 50,000 BCE, the Poseid Islands became a continent and the Pfrees of Mu were now Atlanteans. Their extreme emphasis on technology became the advent of patriarchy — it was science and wealth for its own sake with the spiritual and female aspects submerged. Enough of Mukulian influence survived for a while to effect the growing Atlantis culture, but the patriarchy won in the end. The changeover from matriarchy to patriarchy that women are aware of may have begun with Atlantis.

But initially, this colony of Mu in the Atlantic grew and flourished, and its developing technology began as ways to make life more pleasant for its people. From 50,000 BCE until the earth changes of 26,000 BCE, Atlantis reached its height and fell; an influx of citizenship-trained people from Mu initially helped it develop, but Atlantis went its own way. Technology had been discouraged on Mu, and now the former Pfrees let their imaginations out. From a disgraced colony of the Motherland Mu, Atlantis became a technological wonder:

> The Atlantean scientific turn of mind combined with their energy and aggressiveness made possible great strides forward in mechanics, chemistry, physics and psychology; for in many ways the Atlanteans were a superior people. In certain fields they were more learned than any civilization since.[13]

Atlantis had electricity, heating and lighting systems. Atomic power was made from uranium and used in transportation and in lifting heavy objects; it was abused as weaponry later. There was communication with other lands, and boats, aircraft and submarines. There were light rays, used in healing as lasers but also later used as a "death ray," and a knowledge of anti-gravity. There were metals and metal alloys, some of them no longer known. Rubber was manufactured and there were other factories. Telephones and elevators existed, as did photography and telescopes, and even a form of television. With the dinosaurs gone, cities were now built above ground out of stone, though underground living also continued. Poseidia, Amaki and Achaei were the names of three cities, and these had viaduct systems to bring water to the buildings. For the first time, there were also armies and law enforcement, and the government became a monarchy. Most things began positively, but some were later abused as the patriarchal mindset took hold. Spirituality was important in the early days, brought from Mu and mostly woman-centered. Since all cultures until 5,000 years or so ago were Goddess-based, Atlantis must have been, too. The continent held some of the first great Goddess temples and healing places.

Crystal technology was developd in Atlantis and reached heights never equaled anywhere else again. The Firestone or Tuaoi Stone was developed initially as a way of enhancing the connection between the physical and spiritual, of bringing Goddess cosmos awareness to embodied people. The great crystal was a solar generator and the major power source of the nation. Energy from the sun was reflected and amplified through the crystal, and then stored and transmitted throughout Atlantis. Its energy or electricity was used to power ships, aircraft and other vehicles. Crystal energy was used for laser healing, and for electrical power in the homes.[14]

Helm, speaking through Mari Aleva, describes work with the crystal in the early days of Atlantis:

> There were ancient times when women were more, there was the culture where we had, ah, don't know the words, but we had a big board with different lights and switches on it, matrices and different things and women would run this and some of the men would, too. Where they did not care about which sex they were, but who was astute in the head, who was good in the head. And the ones who wanted to learn and were good at what they do, they would put them in a position that was best for that person regardless of their color or whatever.
>
> Diane: What did the matrix boards do? What were they for?
>
> Helm: Oh, powers of the crystals, of course. And they would light and they would do things. There were energy boards.
>
> Diane: And this was in what we would call Atlantis?
>
> Helm: Oh yes, it was.
>
> Diane: They were an alternate dimension from earth, rather than being on earth?

Helm: Both.[15]

Crystal and gemstone energy were used for healing, as was aromatherapy, as well as a number of the healing skills women are rediscovering today. Healing with herbs and essences would have been known, as well as midwifery, body work, touch healing and perhaps Reiki. Helm also describes what today would be called monasteries, where women and some men went for learning. While women were still dominant, some men were involved with spirituality:

Diane: So the women were basically dominant and the men were protectors?

Helm: There have been many cultures, the culture that we can share with you does have what you are saying. The women were leaders in those communities.

Diane: And that was in the communities that we call Atlantis, but were other dimensions rather than the earthplane as we know it?

Helm: We see there were some in male bodies who were very wise ones and were what you would call in this time, not convent, but in Tibet they have them, and what are they called?

Diane: Monasteries.

Helm: Monasteries, type of thing, where they go in there and they work with wisdoms and they work their own inner magic. The men would do this and the women would do this, too.

Diane: Was it women then that abused the technology? And caused the end of it all?

Helm: The technology is another story. There were not only the people from one country involved here. There were others wanting what some people have, other one want to have it, and there would be destruction rather than give it away. Does this sound familiar?[16]

In this section, Helm talks about a pole shift. I suspect that her information on the monasteries, and a culture where women were dominant was Mu or very early Atlantis, as such equality was eroded quickly in the later culture. The pole shift that destroyed Mu occurred because of an accident of Atlantean technology, an attempt to destroy the dinosaurs. Another occurred at 26,000 BCE and destroyed much of Atlantis and the last of Mu, as well. Helm emphasizes again and again, as do other channelers, that early earth was settled from other dimensions or planets. See Marion Webb-Former's material below.

The oppression of people seems to have begun in Atlantis, as it drew away from the spiritual, female, Goddess culture of Mu. The half-animal, half-human Be-ings are mentioned again. They were used as slaves in Atlantis, but have not been mentioned in any of the material on Mu, though they must have originated there. Barbara Hand Clow defines these as Be-ings possibly resulting from the misuse of thought power; their nonspiritual natures made manifest in their chosen bodies:

The beings who are denser than the humans are a mixture of animal and human forms. Some of them are gross with deformed limbs and faces, and they can be very strong and violent. Some are part goat, or with dog bodies, or lizard bodies, and some are half sea creatures. Some are very beautiful like unicorns and mermaids. We never underestimate them. I wonder if they are mutations?

No, that is not all it is. It is more like when I go to cocktail parties today, and I see people as their animal nature when I'm bored. I've looked around a formal table and seen pigs, chickens, lizards, and cows, and what I was seeing was a facet of their personalities.[17]

Ruth Montgomery and Edgar Cayce called the mating of humans with animals in the early embodiment on earth original sin.

Whatever or whoever they were, they seem to have been mistreated in Atlantis, considered machines rather than as living creatures. Both Ruth Montgomery and Edgar Cayce call them "things." One activity of the healing temples was to attempt to change these Be-ings into fully human form. The Temple of Sacrifice and the Temple of Beauty in Atlantis were healing centers dedicated to this. Sources describe the removal and reshaping of "appendages" (wings, claws, nonhuman attributes) done in these temples, apparently by surgical means. There were attempts to make these Be-ings emotionally human as well. But they were an oppressed class that eventually died out. They are mentioned in Sumer, but don't seem to have survived the destruction of Atlantis.

Edgar Cayce also describes Atlantis as being dominated by red-skinned people, whereas Mu describes the development of all the races and their equality. Perhaps the early tribe of Pfrees that left Mu for Atlantis was a red-skinned group. As Atlantis moved further from Mukulian spirituality and values, they seem to have abused other races as well. Automatic writer Marion Webb-Former channeled this information:

Throughout their history, Atlanteans did little to expand their territories. During the earlier years of their existence, they had reconnoitered their planet and gleaned whatever they required, which included bringing many from the white and yellow races to their continent. These captives were used as slaves and were considered considerably inferior to the Atlanteans and there was no intermingling of the races on Atlantis. They quickly turned their interest to the stars and developed a technology which would make the nearby planets accessible to them. Man will find remnants of their existence on Mars, Jupiter and Saturn.[18]

Mu had sent out emigrees to China, the Gobi Desert, India, South America, and other places; some members of these tribes or families were later taken captive to Atlantis. Giving an extraterrestrial origin for the people of both Atlantis and Mu, Marion Webb-Former indicates that the connection with other planets continued during this period, but did not survive the second earth changes of 26,000 BCE. Atlantis used aggression in attempt to expand its territories, particularly in its last period, but failed.

Mu developed in the "harm none" mode, with values that are typical of those in other Goddess cultures and in those where women and women's values are dominant. Atlantis diverged from that model, becoming a patriarchy — the Pfrees were so disruptive that Mu had to ask them to leave. Where the rights and respect for other races declines, the rights of women also decline. What happened in early Atlantis after the destruction of Mu was the beginning of the patriarchal order women know today. With the devaluing of women comes the devaluing of other groups of people outside the ruling class, the advocacy of armies, aggression and wars, the manipulation of individuals and the abuse of technological advances. All of these are evident in the history of Atlantis, and all are only too evident in the history of the world today; Atlantis has often been compared to the United States. The women's New Age we are now moving toward will hopefully turn the order around again. A psychic reading by Tanith put it this way:

> They came in two forms, female and male, both with promise, both with potential, but both pulled off the true path and track. Where woman needed to create and celebrate life, she learned to fear death and invasion. Where man needed to protect and defend he destroyed and bullied and battered and beat and abused. As they grew further apart, the harmony of the world grew further into chaos, further into disorder and further from harmony. The fire raged, the water wept, and the earth felt pain, for the air did not know harmony. [18]

Atlantis was not entirely corrupt; many with old values lived there. Its knowledge of healing is something women are working to regain today, and much of its science was positive and is still lost. There seems to have been a great tension between those working toward spirituality, equality, harming none, and the positive uses of advanced technology, as opposed to those whose motives were less pure, greedier and clearly manipulative, as there is in our culture today. The earlier days of Atlantis, when the influence of Mu was still strong, were usually more positive than later, when patriarchy was fully developed, morality had declined and the technology was being systematically misused. The tension between philosophies seems to have been a pervading fact of Atlantean life. Says Marion Webb-Former, comparing Atlantis and Mu, and writing of the early earth experiment:

> Even before true man was placed upon Earth, the older cultures of original man and the man-beasts were being watched and influenced by...inhabitants of what man calls "the dog star"....The people of Lemuria mostly heeded the teachings of the good watchers, which advocated the oneness of all things; whereas the people of Atlantis preferred to pursue the glorification of the individual in ways which were detrimental to the uniformity of the whole life force. [20]

In the experiment of the creation of people and their settlement on earth, other planets were clearly involved. When humans grew greedy, abusive and patriarchal, earth was quarantined from contact with the rest of the universe. One of the things coming in a women's New Age is a return to connection with other planets. Most people are already aware that the UFOs are real. There has always been a tension between good and evil in the affairs of people; and there has always been free will to choose. By the time of the second earth change in about 26,000 BCE, the height of Atlantean culture, morally and scientifically, was past.

That second earth change, also possibly accompanied by a pole shift, resulted in the final sinking of the remnants of Mu and the breaking up of Atlantis into two large islands. The cause of the devastation may have been the great Firestone, the great crystal, that was accidentally tuned too high. Marion Webb-Former in her channeled writing indicates that a civil war may have been the more direct cause; the Firestone or other technological destructions used in it were powerful enough to destabilize the earth. Modern women have speculated on the possibility of similar destructions in the advent of a nuclear war. Marion Webb-Former writes of the second earth change and the destruction of most of Atlantis:

> When Atlantean society had peaked and within its decline was creating great suffering for certain of its classes, the technological basis of the weaponry used during the resulting civil war caused a geologic chain reaction which virtually disintegrated the eatern portion of the land and split the western section into two. Thus all that remained of Atlantis was two islands. The larger of these was located off the coast of what is now known as Portugal and it continued to exist for a further twelve thousand years. At that time (approximately 10,300 BCE) there was a slight axis shift, created by solar activity, and the larger island sank and the smaller island broke apart, but remained above the waves. Therefore, there is a vestige of Atlantis still remaining in what you term the present time and it is known as the British Isles...[21]

The third and most recent earth change that destroyed the last remnants of Atlantis, occurring about 10,300 BCE, is the one written about by Plato, who got his information from Egypt. That destruction was also felt worldwide as the great deluge or great flood. More on it in the next chapter. Marion continues:

> At the time of the second devastation...(26,000 BCE) its civilization had, as happens with all cultures, already lived through its full potential and, after that...catastrophe, much of its former technology was lost. Earthly Atlanteans were no longer able to have any form of contact with those who had settled on other planets. Some of those who did not perish shared what remained of their continent with others..., others transported themselves to what is now known as South America, which was considerably larger then, and Antarctica, which was not under the ice flow at that time. Here they mingled their culture with the descendents of the

refugees from the devastation, long before, of Lemuria. [22]

Ancient maps, including the Piri Reis map that Columbus would have known, show Antarctica without its covering of ice. The maps display geographic information only rediscovered in the 1950s by sonic soundings. Some of these old maps also show Atlantis, or where it had been. Refugees from Atlantis went to Antarctica and to join the Mukulians: at some unknown point in South American history, the gentle Incas were taken over and dominated by a less peaceful people. Were these Atlanteans?

Speaking through Mari Aleva in her Eastern European accent, Helm describes to myself and Caridwyn Aleva this second devastation and the meaning of a pole shift. Science believes that a number of such shifts have occurred in the evolution of the earth:

Helm: There were eruptions of lands. There were shifts of land, from pull apart from different...

Diane: A pole shift?

Helm: Poles would do magnetics, has to do with sky and with your gravitational pulls of energies bypass around and it would combust. We have a barrier here with language.

Caridwyn: Helm, does this relate at all to the land break between Africa and South America where the pieces fit together if you move them together, but they began to drift apart after a major rupture?

Helm: You understand. I want to tell Madelein (Mari) something. First for you. Understand of land erosion, lot of erosion here.

Diane: Wearing away?

Helm: Now to tell, I want to speak with Madelein here. She say how can land float? Let us explain, honey. You have your core, and you have a liquid mass here. There is metallics in the liquid mass and there is your polars. Then as it cool you have your crust. And as crust splits apart, it is still connected to its core and you have what you call water which is liquid and can move. The land as things move can move when it is connected to a core. Do you understand?

Diane: Yes.

Caridwyn: These are the tectonic plates that they talk about.

Helm: Well, Madelein here does not understand. This is only way we can explain, connected and they can move.

Caridwyn: It's as if you had the center of a wheel with the spokes moving out and if some of the spokes start moving apart, they're still connected to the center though there may be wider spaces.

Helm: Absolutely. That is right....You see Madelein think that you talk about island, that is like a piece of ice that float on top of water.

Caridwyn: No, it goes down deep....

Diane: So, this was the pole shift that destroyed Atlantis?
Helm: As the pole shift, think of her wheel here and here and as things move they wiggle. Ah, and if it move here, it is still connected and the water in between erodes. The cultures come back and forth around from once walk on land and they say, how does this happen?[23]

Helm says there is an Atlantean ship under the ocean that will be discovered one day — a spaceship. Sumerian texts also talk of a spaceship accident and a ship lost under the sea. Are the cultures connected? Channelers tracing human origins on earth speak of the influence of several different or planetary systems.

After this destruction, Atlantis' power was mostly broken, but it still retained superior technology that it tried to preserve. Atlantis sent out colonies, knowing that their remaining land was geologically unstable, and this is the period of Atlantean aggression in Europe and South America. The war between Atlantis and India probably happened at this time.

When the Pfrees that became Atlantis left Mu, the group of Katholis that became India and the Rama Empire also left. The two groups had been opposing factions, representatives of two extremes, and their enmity continued. The thousands-of-years-old records of Atlantis' attack on India are contained in two ancient manuscripts, the *Ramayana* and the *Mahabarata*. While it is hard to date the actual occurrence of this war, I believe it to have been either in the time just prior to the 26,000 BCE earth change or not long after it, in the period when Atlantis was seeking to gain a foothold in other lands.

India won the war over its Atlantean attackers, but the Atlanteans came back after the defeat with nuclear weapons. The descriptions of warfare in the Indian epics are only beginning to be understood today; until 1945 such warfare was not considered possible. After the final end of Atlantis, Atlantean technology was lost and much of it is only now being rediscovered. Unfortunately nuclear power has been relearned, and used to destroy cities. Channelers today believe that the threat of a nuclear war destroying modern civilization is over: the example of Atlantis has also been relearned, at least that far.

Atlantis at the time of the war with India was "a highly technical, patriarchal and war-like culture...bent on destroying the world."[24] It had aircraft, rockets and the knowledge of anti-gravity, along with other advanced weaponry, and it was the aggressor. When Atlantis attacked the Ramas, India used mind power to stop the war by causing the leaders of the Atlantean army to die of heart failure. The Atlanteans withdrew, then sent devastation. The following is quoted from the *Mahabarata:*

Gurkha, flying a swift and powerful vimana,
hurled a single projectile
charged with all the power of the Universe.

An incandescent column of smoke and flame,
as bright as ten thousand suns,
rose with all its splendor.

It was an unknown weapon,
an iron thunderbolt,
a gigantic messenger of death,
which reduced to ashes
the entire race of the Vrishnis and the Andhakas.

The corpses were so burned
as to be unrecognizable.
Hair and nails fell out;
pottery broke without apparent cause,
and the birds turned white.

...After a few hours
all foodstuffs were affected...
to escape from this fire
the soldiers threw themselves in streams
to wash themselves and their equipment.[25]

Archeological evidence for this devastation has been discovered at the
Indian sites of Harappa and Mohenjo Daro — the Vrishni and Andhaka of the
epic. In naming the destruction of these cities, the *Ramayana* (in quotes) and the
Lemurian Fellowship write of three ages:

"We now live in the third age of Time, and Rama lived in the second age of the
world." Perhaps the time of Mu, the supposed Mother-civilization of the world was
the first age, the time of Atlantis and Rama the second age, as the *Ramayana* says;
the time after the destruction of Atlantis and Rama and the time when the great
epics were written was the third age, and the fourth age is the time in which we
are now living.[26]

After the earth changes of 26,000 BCE, survivors sought to preserve the
learning of Atlantis, knowing that the two remaining land masses were unstable.
Says Marion Webb-Former:

During the last twelve thousand years of what remained of Atlantis, its scientific
class endeavored to preserve what remained of Atlantean technology and also
attempted to regain some of the knowledge which had been lost. As the millennia
slowly passed, they became more and more involved in seeking ways to secure the
safety of their technical heritage. They knew the land on which they dwelt was
extremely unstable and they were fearful that one day all trace of Atlantis would be
gone forever, taking with it all its ancient wisdoms and learning.[27]

Emissaries were sent to find a nation or culture that would be the recipient of Atlantean learning, and a tribe of people living in the Nile basin was chosen. With Atlantean training and influence, these people became the first beginnings of ancient Egypt. They and India got from Atlantis, via Mu, their creation story of the cosmic egg and the Goddess that was later changed to male in both cultures. Egypt was a center of civilization, influencing the cultures of Africa, and Greece via Crete. Crete may also have been an Atlantean colony, or settled by survivors of Atlantis. The Egyptian pyramids were Atlantean-built, centers of the technology and learning brought to a new land. According to several psychic sources, a long-hidden library, containing the full stories of Atlantis and Mu plus other information, will one day be rediscovered in an underground chamber connected to the Great Pyramid. Egypt became the center of the new world, the fourth world, the world that began when Atlantis died.

In the last thousand years of Atlantis, most of the remaining culture and inhabitants were housed on the larger island, as it was believed that the smaller island would disintegrate first. Ironically, it was the opposite, and when the larger island sank, only the knowledge of the people who had colonized elsewhere remained of the Atlantean empire and culture. Again from Marion Webb-Former's channeling:

> The smaller island was mostly devoted to rural living and any technological industries were moved to the larger island. It was feared that the smaller island would disappear first, therefore, by the time of the final catastrophe, it was home to only those who wished to live a very simple existence, gleaning what they could from the land and pursuing their belief of WHAT IS in whatever way they chose. Their society became a contradiction of itself, being both practical yet mystical. Gradually, they had less and less contact with the larger island and, after the last axis shift, their break from the old Atlantean ways became complete. Much of their ancient knowledge was lost to them and slowly they regressed into a very primitive race. Only within certain of their religious ceremonies and beliefs could the old learning still be perceived, yet not usually understood. Stonehenge is an example of a physical remnant from Atlantean life before the final devastation. [28]

That smaller island, as Marion has stated in her writing, is now the British Isles. Much of western wicce, the Old Religion, comes from here.

Atlantis and Mu, the patriarchal and matriarchal, influenced and became all the nations of the earth. The cultures of the South Pacific and Central and South America, as well as Mexico, China, India, Tibet and western North America are believed to be survivors of matriarchal Mu. The cultures of Britain, Ireland, Spain, Portugal, Egypt, Crete and Greece and possibly the early African empire were Atlantean colonies or remains. The Mound Builders of native North America were probably Atlantean — Frank Waters' *Book of the Hopi* includes them in the Hopi creation epic. The Hopi legends describe the four worlds with a story of the three earth changes that devastated Mu and Atlantis. They describe how the Pueblo people settled South, Central and southwestern

North America after the last destruction of the world. The records hidden in Egypt will be recovered one day, and the stories of Atlantis and Mu, and their origins from the stars will be revealed.

Modern western science totally denies the existence of Mu and Atlantis, though documented archeological, linguistic, geographical and geological evidence makes the patriarchy's stance ridiculous. Explorers of the sea floor of both the Atlantic and Pacific have found remains of the ancient cultures and their advanced civilizations. So much exists that can only be explained by the concept of lost continents that to ignore them is a breech of science's own methods. Is it the matriarchal, Motherland roots they deny, or the dubious history of their own patriarchies?

I fully believe that the changeover from matriarchy to patriarchy happened in Atlantis, and the early Goddess golden age was the Pacific continent of Mu. Time and the stance of history, which claims all cultural advances as men's alone, has kept herstory from women. Since the excess of Atlantis continued after the continent's destruction, herstory has continued to be suppressed. The changeover from Goddess to the male god, and the destruction of women's early culture and writings, coupled with the takeover of Goddess cultures everywhere has completed the coverup and loss.

Mu was the first world, a Goddess world, and Atlantis was the second. The time of the *Ramayana* was the third age, the time when the Goddess was suppressed and patriarchy conquered the remaining continents. We of today are the fourth world, about to enter the New or fifth age. Who will we choose to emulate — the peaceful, matriarchal, spiritual people of Mu, or the warlike, patriarchal, technology-for-its-own-sake Atlanteans? New Agers today compare Atlantis with the United States, and believe that many people incarnated now on earth were once incarnated in Atlantis. Those incarnated today were on both sides of the early Atlantean struggle; they were those who "harmed none" and those who corrupted the culture with greed and power-over. Again today we have a choice of how we want to live, what we want to create of our world. It's a choice to survive and blossom, or a choice to die and be submerged. It's a choice of matriarchal peace or patriarchal war and destruction. Which will women choose in the New Age?

Notes

1. Barbara Hand Clow, *Eye of the Centaur* (St. Paul: Llewellyn Publications, 1986), p. 182.
2. Edgar Evans Cayce, *Edgar Cayce on Atlantis* (New York: Warner Books, 1968), pp. 49, 77.
3. Col. James Churchward, *The Lost Continent of Mu* (Albuquerque, BE Books, 1959), pp. 21 ff.
4. *Ibid.,* pp. 28-29.
5. David Hatcher Childress, *Lost Cities of Ancient Lemuria and the Pacific* (Stelle, IL: Adventures Unlimited Press, 1988), p. 28.

6. Lytle Robinson, *Edgar Cayce's Story of the Origin and Destiny of Man* (New York: Berkeley Medallion Books, 1972), pp. 54-57.

7. David Hatcher Childress, *Lost Cities of Ancient Lemuria and the Pacific*, p. 28.

8. *Ibid.*, pp. 344-347.

9. *Ibid.*, pp. 28-29.

10. *Ibid.*, p. 29.

11. Mari Aleva, *Earth Changes with Mari Channeling*, See appendix, p. 181.

12. David Hatcher Childress, *Lost Cities of Ancient Lemuria and the Pacific*, p. 351.

13. Lytle Robinson, *Edgar Cayce's Story of the Origin and Destiny of Man*, p. 58.

14. *Ibid.*, pp. 58-59.

15. Mari Aleva, *Earth Changes with Mari Channeling*, appendix, p.192.

16. *Ibid.*, pp. 181.

17. Barbara Hand Clow, *Eye of the Centaur*, pp. 151-152.

18. Marion Webb-Former, *On Atlantis*, p. 2, unpublished.

19. Tanith, *Reading, May 7, 1990*, p. 2, unpublished.

20. Marion Webb-Former, *On Atlantis*, p. 1.

21. *Ibid.*, p. 2.

22. *Ibid.*, pp. 2-3.

23. Mari Aleva, *Earth Changes with Mari Channeling*, pp. 181-182.

24. David Hatcher Childress, *Lost Cities of Ancient Lemuria and the Pacific*, p. 71.

25. Quoted by David Hatcher Childress, pp. 72-73.

26. *Ibid.*, p. 75.

27. Marion Webb-Former, *On Atlantis*, p. 3.

28. *Ibid.*

More Ancient Herstory
The Legacy of Mu and Atlantis

Male bias, together with preconceived religious attitudes, which appears in both major and minor matters, raises some very pressing and pertinent questions concerning the objectivity of the analysis of the archeological and historical material available at present. It suggests that long-accepted theories and conclusions must be re-examined, re-evaluated and where indicated by the actual evidence, revised.

Merlin Stone[1]

After the second earth change of 26,000 BCE, Atlantis no longer had connection to other planets. The continent broke into two islands, its height of civilization past, and the last remains of Mu were also lost. Human souls, those Be-ings that first embodied on Mu and developed the foundations for Goddess matriarchy, were now on their own, struggling to continue civilization under patriarchy's advance. What happened to the survivors of the sinking of Mu and Atlantis, where did they go? What portions of Mukulian and Atlantean culture, technology and religion survived? What happened to the matriarchal peoples brought to earth from the stars?

As has already been traced, survivors from the 50,000 BCE destruction of Mu went primarily to South America and to Atlantis. Colonies of Mu already were established in China, India, Central America and Tibet. Remains of Mukulia today include the Polynesian and Pacific Islands, Australia and New Zealand, and western North America. Atlantis' migrations went to Central and South America, and also to the east. Around 11,000 BCE Atlantean knowledge developed the high culture of ancient Egypt, whose influence spread through Africa and later into classical sources of western civilization — Crete, Greece and Rome. England, Ireland and the Celts were survivors of the last remains of Atlantis, as were the Canary Islanders, and the Basques of the Spanish Pyrenees. Sumer also changed at this time: First established with help from the Nefilim, the Sumerians were then left to develop alone. All of these cultures, as geographically separate as they are, have records of the great flood of 10,300 BCE, of their civilizations before it, and of rebuilding after.

Egypt was the culture chosen by Atlantis to carry on Atlantean civilization. The teaching process was never completed, and after the end of Atlantis, Egypt mingled Atlantean teaching with its own ways. The changeover in Egypt from a Goddess pantheon to a god-based society (Ra, the sun; sun worship was brought from Atlantis) happened during the time of Atlantis' influence on early Egypt. Yet, the great Goddess Isis, whose name is connected with Sirius or Sothis, the Dog Star, became the mother of the developing world internationally for thousands of years after Atlantis ended. Isis was Demeter in Greece and Crete, and Emme Ya (Yemaya) in Africa. She may have been the name of a very early, pre-Atlantean queen — or the name of a planet of origin. The patriarchal influence of sun god Ra, who actually was an Atlantean priest-teacher, initiated the Egyptian sun-worshipping dynasties.

Information on Egypt for this chapter comes mainly from Edgar Cayce and the channeled writings of Marion Webb-Former. Marion channels an entity, Moon, who was present at the time of Atlantean influence in Egypt as Tefutti, or Thoth, another Atlantean teacher. Marion herself was Mayatta at this time, an Atlantean priestess and wife of Tefutti. The disparities between the scientific ruling class of Atlantis, and the less-respected but still highly placed priest/ess class are apparent in her writing. The tension between science and morality seems intrinsic in the two classes' roles. Ra, Tefutti/Thoth's scientist brother, was also known as Ratta. Those who participated in the project of bringing Atlantean advances to the Nile seem to have had better motives than the majority of the scientific-patriarchal rulers of Atlantis. This was the last ending of Atlantis, when the results of technology-for-its-own-sake were probably only too apparent. The time was about 11,000 BCE.

Who were the people Atlantis chose as its successor? Marion and Moon write:

> It was decided that a small race of primitive people, living within the area of what is presently called the Nile basin, would be the best choice. These people were the descendants of a group of marauding, white nomads who had finally settled in this fertile region....The scientists decided that these people were obviously aggressive enough to pursue the rigors of Atlantean existence and sufficiently intelligent to be slowly taught all the beliefs and knowledge of their dying culture. [2]

Tefutti/Thoth, a member of the scientist ruling class, was sent as an envoy to these people. Since he traveled in a flying machine and did things the natives could only perceive as magic, he and the other Atlanteans after him were looked upon as gods. These advanced people were tall and long-lived:

> The people who lived on Mu and Atlantis lived for many hundreds of years, as did all men when they were first created, but only those who kept alive the wisdom of what the physical universe truly is were able to perpetuate long existences within that universe. Thus, those who had settled in the Nile basin, as with most tribes existing at that time, lived very short lives compared to the Atlanteans....these people looked upon me as a god for I was able to live with them, leave, and then return many, many years later, and surely only a deity possessed such powers. [3]

Ratta, or Ra, later replaced Tefutti in Egypt. Ratta was a priest rather than a scientist, and aware of the wrongs of his culture — but on a mission. While working together again, the two brothers often clashed — Tefutti/Thoth, developer and teacher of language and science, was loved but seen as a demiurge — while Ratta was accepted as ruler and became the Egyptian god. Ra in Egypt was the sun, and sun worship seems to have developed in Atlantis. The change from Goddess to god in archeology is often accompanied by the change of worship from the female moon to the male sun. Yet, says archeologist Patricia Monaghan, most sun gods were originally sun Goddesses. If the Mukulian creation story speaks of the serpent in the egg and the mingling of the sun's rays with the earth, it is possible that the first divinity was a sun Goddess. She may have been changed to male by the increasingly patriarchal Atlanteans. When the Egyptians saw the technologically advanced Atlantean people, they thought them gods, taking their nature religion mixed with Atlantean mysticism down other paths entirely.

According to Edger Cayce, Ra-Ta (as he spells it), was the first to establish patriarchal marriage in Egypt, insisting that both men and women be monogamous; marriages had to be approved by the king, whose dynastic line was also chosen by Ra. Ra segregated men from women in separate sleeping quarters. Children, who were delivered by priestesses in the temples, were separated from their mothers at three months and reared for the state by professionals. There were separate temples where matings were allowed, but only between paired couples. Some of the partners were Atlantean, and the motive was genetic upgrading.[4]

The Atlanteans initiated commerce and mining, stoneworking and banking, literacy and libraries. Ra established the country as a monarchy that included a civil service and some facets of representation. He taught the ideas of the immortality of the soul, reincarnation, karma, and the soul's beginnings and evolution in and out of bodies.[5] Development was slow by Atlantean standards, so the pyramids were built as places of initiation and learning centers. The combination of the pyramid shape and the crystals used in their capstones was a design for raising the spiritual awareness of a chosen Egyptian priesthood (some women, but predominantly male). A number of pyramids were built, though only two remain, and they are much older than is currently believed. The design was from Atlantis, and storehouses of knowledge were hidden under and around them.

Atlantean builders, using their advanced technology, completed this task and it was accomplished with ease and swiftness....Because...Ratta was a high priest, he knew many ways to teach an earth-bound soul of what its true nature was and, because I (Thoth) was of the scientist class, I was cognizant of methods to enhance the human brain. Thus, together, we copied and improved upon those indoctrination rituals which were reserved for only an exclusive class of Atlanteans. We needed a rapid understanding of all the Atlantean knowledge which had been amassed over thousands of years and, with the help of pyramids and crystals, we began to accomplish our goal.

Within only a few hundred years the race of people within the Nile basin advdanced tremendously. Cities were built under the guidance and direction of many Atlanteans, some of whom remained in this new land. The mingling of those of the red race with those of the white race created individuals who are considered by the modern world as ancient Egyptians. Their culture was created from their previous nomadic existence and their more recent exposure to Atlantean life.[6]

Egypt, which became a center for the rest of world civilization to follow, was trained by the Atlanteans, although some of that training became confused along the way. In similar ways, the people of Sumer were guided by the advanced extraterrestrials from Marduk, who may have genetically created them. Sumer's connection, if any, with Atlantis and Mu is not clear, but Sumer is also considered a root culture of earthly civilization. The influence of other planets seems to be a recurring thread in human evolution and development.

The great flood is usually believed to have caused the last sinking of Atlantis at 10,300 BCE. All of the cultures believed to derived from Atlantis and Mu — and then some — have a tradition of the flood, complete with their own versions of how people, or at least those who had divine help, survived. The story in Sumer continues the narrative in ancient texts of the Nefilim and the humans they created.

According to Zecharia Sitchin, whose book *The 12th Planet* provided the history of Marduk and Tiamat, the people the Nefilim originally made were designed as drones or dependent workers. They were possibly hybrids (of *Homo erectus* and the "gods"), not equipped with reproductive capability, which was given or discovered later. Eve's eating of the Tree of Knowledge, defined by some scholars as knowledge of sexuality, and by others as knowledge/ awareness of Be-ing, was the dividing point between people as workers-only and people as autonomous humans. The capability to reproduce seems to be a part of this awareness, which is another explanation of original sin. Once the capability of awareness/knowledge evolved or was given scientifically to these early Adamas, the ability of procreation came with it, and the Be-ings were now fully human. Woman seems to have become human first!

Sitchin does not define awareness, reproduction or the acquiring of either as original sin, but considers the fact that once humans could reproduce they were genetically compatible with the "gods" of Marduk. Their unions with each other were said to weaken the genetic lines of *both* species, and the leaders of the Nefilim were displeased.[7] Is this another interpretation of the "pure souls" merging with animals to create a race of half-humans, half-beasts? Genetics seems to have been an important issue for the early developers of people. The Nefilim were deliberately breeding workers; the Atlanteans controlled Egyptian reproduction to strengthen them genetically, and the older, somewhat confused story of the half-people, half-animals appears in several creation legends. Are these genetic rules and prohibitions simple racism, or were there legitimate reasons for it? On matriarchal Mu, all people were considered equal; superiority came through education and training that was available to all.

At any rate, in Sumer's story as translated and interpreted by Sitchin, Enki was the genetic engineer of Adama, and was responsible for making Adama fertile later. When a council of the gods decided to destroy all of the humans, supposedly to keep their own race "pure," Enki secretly warned Utnapishtim, the human administrator of the Nefilim/Sumer city of Shuruppak, to build a boat. His act was one of defiance, and unknown to the rest of the gods. [8] The launching of a Nefilim rocket was the signal to launch the ark, to take aboard all the humans and animals to be saved. Utnapishtim (in other versions called Ziusudra, Atra-Hasis or Noah) was chosen to survive because his line had not mingled with the Nefilim, but was genetically fully human. The rocket taking off was the Nefilim themselves fleeing the beginning destruction. [9]

Ninti and Ninhursag, Birth Mothers to the new race, reacted to the deluge from their escaping spacecraft:

> Ninti wept and spent her emotion;
> she wept and eased her feelings.
> The gods wept with her for the land.
> She was overcome with grief....
>
> The Goddess (Ninhursag) saw and she wept...
> her lips were covered with feverishness....
> 'My creatures have become like the flies —
> they filled the rivers like dragonflies,
> their (mother)hood was taken by the rolling sea.' [10]

Sitchin gives a variety of possible reasons for the deluge itself, and the Nefilim's knowledge of when to leave earth and abandon their creations to destruction. Among the suggested causes are the rising of the oceans to tidal wave levels due to the sudden melting and slippage of the Antarctic ice cap, or a pole shift created by the passing close to earth of the 12th Planet Marduk itself. (Most sources describe Antarctica as temperate until the sinking of Mu at 50,000 BCE. Mu attempted to colonize there, but the changed weather became uninhabitable.) Either of these could have also resulted in the final destruction of Atlantis, or the sinking of Atlantis could have caused the deluge. Whatever happened, the scientifically advanced Nefilim had warning. Sitchin dates the first landing of Nefilim on earth at 445,000 years ago, the fashioning of people by Enki and Ninhursag at 300,000 years ago, the merging of the races of Nefilim and people at 100,000 years ago, and the great flood at about 13,000 years ago. [11] The time of the final sinking of Atlantis was approximately 10,300

BCE — about 13,000 years ago.

When the Nefilim returned to earth after the deluge and discovered that some humans had survived, they vowed to help them. The sudden development of civilization in the Near East around 11,000 BCE was the result, and the Nefilim left the earth permanently sometime after. The loss of contact with extraterrestrials may have happened in Sumer later than in Atlantis. According to Ruth Montgomery, the original people of Sumer were an early Mukulian colony, and the later-landing Nefilim were Atlanteans.[12] Sitchin's story of Sumer is taken completely from the culture's own writings, and to me seems closer to the truth.

A later outgrowth of Sumer was Babylonia, and the Babylonians describe the founder of life on earth as an amphibious creature named Oannes. Oannes in art is pictured as standing upright, and being part human and part fish. Likewise, the Egyptian Goddess Isis is always drawn with what looks like a small fishtail in her headdress or hair ornament. Both Oannes and Isis are connected to the planetary system of Sirius, the Dog Star, and are said to come from there. Another name for Isis is Sothis, the Egyptian name for Sirius. Oannes may be another name for Enki.

In Mali in sub-Saharan Africa, quite far from Egypt or Sumer, a small tribe of Black people called the Dogon describe a similar creature to Oannes as the founder of their religion and culture. They call the Be-ing the Nommo, and say it came to civilize and teach them, landing in a spaceship. Not far southwest from where the Dogon live are a people who call their creation Goddess Yemaya, and describe her as a mermaid, part woman and part fish. Remember the half-human creatures of other cultures — were they visitors from Sirius?

The Dogon people describe their origins as from this far away star system, and have astronomical information about the Sirius system that is only now being verified by scientific star mapping. They describe a smaller star near Sirius that they call *po tolo,* known as Sirius B to astronomers, and another star in the same elliptical orbit, *Emme ya,* or Sirius C. Scientists are aware of Sirius B, which is as invisible without good telescopes as the Dogon know it to be, but they have not yet verified Sirius C. The Dogon describe the orbits and movements of these stars in ways that science can only guess the accuracy of. The Dogon creation story says that the Nommo came to them from Sirius C, the tiny Emme ya that they also call the Sun of Women, and a planet that circles Emma ya.[13] Neither Emma ya nor her tiny planet have yet been discovered by science. The Dogon are not a scientific people, but they know their cultural roots.

I have changed the creator's pronoun to 'she' in this Dogon account of life on other planets:

> The worlds of spiralling stars were populated universes; for as (she) created things,
> Amma gave the world its shape and its movement and created living creatures.
> There are creatures living in other 'Earths' as well as on our own...[14]

Each star system is called a "placenta," the earth's solar system is called "Ogo's placenta." The system of Sirius is "Nommo's placenta," Nommo being described as half-human and half-fish, an amphibious creature. The Nommo's Sirius system is called "the land of the fish." Ogo is the Fox, who is impure, and who represents both people and our solar system's cosmic impurity. Ogo is imperfect: he rebelled at creation and is an unfinished Be-ing, isolated from the rest of the universe and hoping for redemption.[15]

So much of this sounds familiar. Here is Be-ing birthed from the womb of the Goddess — the placenta is the mother's internal nest that nourishes the developing fetus in mammalian young, and is a female birth symbol like the cosmic egg. From the womb or plancenta is born the "child of the universe" that rebels, and has turned against the plan of creation and remains unfinished and isolated until it grows up enough to learn cosmic connection and oneness. The Nommo is another advanced culture who once helped humans, or brought them here, and has since left us to ourselves to sink or swim, to grow into "harm none" or to destroy ourselves. We are quarantined from others until that decision is made. Sirius, the Dog Star, has been mentioned in a number of creation stories, and there is a planet we can't see yet with our scientific eyes, but only with our spiritual ones.

Here is the Dogon account of the creation of our sun's solar system:

> The landing of Nommo on our Earth is called "the day of the fish," and the planet (she) came from in the Sirius system is known as the (pure) earth of the day of the fish...not (our) impure earth...In our own solar system all the planets emerged from the placenta of our sun. This is said of the planet Jupiter, which "emerged from the blood which fell on the placenta." The planet Venus was also formed from blood which fell on the placenta. (Venus was blood red when she was created, her colour fading progressively.) Mars, too was created from a coagulation of blood. Our solar system is, as we have noted, called the placenta of Ogo, the Fox, who is impure. Our own planet Earth is, significantly, "the place where Ogo's umbilical cord was attached to his placenta...and recalls his first descent." In other words, the Earth is where Ogo "plugged in," as it were, to this system of planets.[16]

Amma is the mother that gave birth to all of the star systems, all of the planets and suns. Birth is clearly female, though the source calls Amma "he." Amma in India is a birth Goddess. The Dogon define the Milky Way as made of placental blood, birth blood, the blood that created the universe and planets. The Yoruba Goddess Yemaya (Emme ya) created all life from the birth waters of her womb.

The Nommo will return, the Dogon say, first heralded by the appearance of a certain star in the sky. The impure Ogo will be "crushed," so all the best of human civilization can emerge and be purified. Ogo is described as "he" and the Nommo castrated him; he is the rebel portion of humanity, the aspect that works against peace and harmony on earth. The symbol of the Nommo's return is *ie pelu tolo,* "the star of the tenth moon," the spaceship. The Nommo is the watcher or monitor of the universe, guardian of the spiritual and owner of the rain and water. The Nommo manifests as the rainbow, called "the path of Nommo," and is the "guardian of the spiritual principles of living creatures on Earth." These attributes are also Yemaya's.

Robert Temple, the author of *The Sirius Mystery* (Destiny Books, 1976), connects the story of the Dogon, the Nommo and the star system Sirius with the mythology/legends of Egypt and Greece. The rising of Sirius/Sothis/Isis at the time of the yearly flooding of the Nile is the basis for the Egyptian calendar, and a central religious cycle, the story of Isis, Osiris and Horus. This yearly cycle became the model of the later story of Persephone and Demeter as well as the still later one of Mary and Jesus. The Dog Star and the Pleiades are the two star systems most mentioned when origins or help from other planets is referred to. These stars set the calendars in a large number of cultures, though they may not be the most visible or prominent objects in the sky.

Oannes seems to have come later than the Nefilim, as the Babylonians came later than Sumer. The Nommo may have come to earth later than the first destruction of Atlantis, or even later than the final sinking and the great flood. The Dogon are possibly a newer people than the Sumerians, and with later extraterrestrial contact. They are probably older than Egypt, and pre-Atlantean Egypt may have descended from the tribal remnants. The Egyptian Goddess Isis may have descended from their tribal remnants. The Egyptian Goddess Isis may have been one of the Nommo, as was Yemaya. Both could have been personifications of the stars of origin (Emme ya and Sirius), the stars from which the Nommo came.

Cultures worldwide share the same Goddesses, giving them different names in different countries. Yemaya in Africa is Isis in Egypt (also in Africa), Kwan Yin in China, Sarasvati in India, Chalchiuhtlique in South America, and of course Tiamat in Sumer. This universal mermaid Goddess, who is a Goddess of water, women and healing and a creation/birth mother, may have origins going back further than Babylonia and Sumer, further than Atlantis, all the way back to Mu and other planets. These birth mothers of earth could also be the Great Goddesses of other planets and stars. All things in the universe are connected and one day we shall know. And one day the Nommo will return.

Another culture that has direct connection with the herstory women never learn in patriarchy is that of the southwest North American Hopi people. The Pueblo Native Americans are believed to have been survivors of the sinking of Mu. Their worldview is one of the oldest on record, and their herstory describes in the creation and destruction of three worlds what may be the destructions of Mu and Atlantis. The Hopi are a matriarchal people whose name means "People of Peace." The years of patriarchy have not been easy on the Hopi; their ceremonies are ending and their culture has been all but lost to poverty, forced assimilation, and the excesses of the United States government and the Navahos.

Like that of Sumer, the Dogon, Atlantis and Mu, the story of the Hopi comes to women through male sources. This is frustrating, as so often in these sources Goddesses are labeled "he" (even when they create placentas, give birth and lay eggs), heras are left out of the tellings, and female cultures are often changed to patriarchies. Sometimes reinterpretation is needed, which I have tried to do as accurately as possible based on the evidence, but much is beyond repair. Most of what is female has been left out of history, and when men write herstory they claim it for their own. Male interviewers seldom talk to women. In this book I have renamed the birth Goddesses "she," and included women as "people" again. No one knows how much of what is female and matriarchal has been lost forever.

The only complete written record of the Hopi was told by Oswald White Bear Fredericks to sociologist Frank Waters. The account is breathtakingly beautiful, but has the feeling that no one asked the women. Waters' *Book of the Hopi* (Ballantine Books, 1963) is carefully and lovingly written, but entirely from the male point of view. One can only wonder which gods were once Goddesses, and what has been changed over the eons. Spider Woman's twins, for example, are male in this account, though female in the Zuni version of the story. I have given this story in my chapter on creation.

The west coast of North America was part of the continent of Mu, and the Hopi are believed to have been survivors of the destructions of Mu and Atlantis. Their story of creation describes the ending of the first, second and third worlds and their emergence into the present fourth world. The Lemurian Fellowship defines the first world as that of Mu, the second as Atlantis, the third as the time of the *Ramayana* (the ancient world of this chapter), and the fourth as the world today. Described as dimensions, channelers say we are now emerging into the fifth world, and the fourth (some say fifth) dimension, which will begin after the next earth changes. My chapter on Atlantis and Mu describes the first two worlds, the third world is the current chapter, and we live in the fourth. The Hopi creation cycle describes how people emerged from creation to today, the fourth world. Their symbol for these emergences is the spiral or labyrinth, that with squared or rounded sides is a Goddess and birth symbol through the world.

The first world was called Tokpela, and in it the Hopi lived a life of joy and well-being. There was no sickness in this world, as humans lived in harmony with the earth and no evil or wrong existed. Tokpela means Endless Space, and the people created by Spider Woman knew oneness with each other, with the animals, and with the earth.

So the First People kept multiplying and spreading over the face of the land and were happy. Although they were of different colors and spoke different languages, they felt as one and understood one another without talking. It was the same with the birds and animals. They all suckled at the breast of their Mother Earth, who gave them her milk of grass, seeds, fruit and corn, and they all felt as one, people and animals.

But gradually there were those who forgot the commands of Sótuknang and Spider Woman to respect their Creator. More and more they used the vibratory centers of their bodies (chakras) solely for earthly purposes, forgetting that their primary purpose was to carry out the plan of creation. [18]

The first world sounds like the descriptions of Mu, where the races and tribes lived together in harmony and there was abundance and peace for all. The second paragraph sounds like the entrapment in bodies described in the evolution of the soul material. With more involvement in bodies and in the material world came more temptation to stray from the plan of connection with the Goddess/cosmos. Early people communicated by thought power.

Lavaíhoya, the Talker, the mockingbird, came among the people and showed them their differences. People separated from each other and from the animals. Káto'ya, the snake with the big head, encouraged the separateness and divisions. Sótuknang appeared to the people who still lived in oneness, informed them that the world would be destroyed, and told them what to do:

Said Sótuknang, "You will go to a certain place. Your *kopavi* (crown chakra) will lead you. This inner wisdom will give you the sight to see a certain cloud, which you will follow by day, and a certain star, which you will follow by night. Take nothing with you. Your journey will not end until the cloud stops and the star stops." [19]

Those whose *kopavis* were still open, who still had spiritual connection to creation/Spider Woman, were able to see the cloud and the star and to follow it.

When the star and cloud stopped, and the people were all gathered, Sótuknang led the Hopi underground to the Ant People, who were to be an example for them. He then destroyed the first world by fire, opening all the volcanos until fire came from everywhere, and nothing was left except the people hidden underground in Mother Earth's womb. The time was 50,000 BCE, the destruction of Mu; channeled sources mention the explosion of gas pockets. The Hopi were comfortable underground: the people of Mu had built underground dwellings as protection from the animals.

Sótuknang created the second world when the earth had cooled, and released the people from the Ant Kivas. The continents and seas were rearranged and the second world was not so beautiful as the first, but it was beautiful enough.

It was a big land, and the people multiplied rapidly, spreading over it to all directions, even to the other side of the world. This did not matter, for they were so close together in spirit they could see and talk to each other from the center on top of the head. Because this door was still open, they felt close to Sótuknang and they sang joyful praises to the Creator, Taiowa. [20]

This was Tokpa, Dark Midnight, the second world. It was not so easy or abundant as the first world. The people no longer lived with the animals, who were wild and ran from them. Survivors of Mu populated all the continents, and they still communicated by thought power. The Hopi began to make things and to barter with others and this began the trouble: "Everything they needed was on this Second World, but they began to want more." [21] If the second world was Atlantis, this "wanting more" definitely has meaning. People wanted material things for their own sake, not because of need. They forgot living in oneness and lived for the goods; they began to quarrel and fight and war began. Remember Mari Aleva's channeling about the second destruction — and how familiar it sounds.

A few of the Hopi remembered oneness, and when Sótuknang had placed them underground again, he destroyed the second world. He commanded the twins (female in the Zuni version of this story) who stayed at the north and south poles to keep the earth rotating properly, to leave their posts:

> The twins had hardly abandoned their stations when the world, with no one to control it, teetered off balance, spun around crazily, then rolled over twice. Mountains plunged into seas with a great splash, seas and lakes sloshed over the land; and as the world spun through cold and lifeless space it froze into solid ice. [22]

The second Hopi world, in destructions that broke up Atlantis and the last of Mu at 26,000 BCE, ended with a pole shift. Survivors who settled on the once-temperate continent of Antarctica found it no longer temperate, and this may also have been the start of the last ice age. Zecharia Sitchin posits the melting at 10,300 BCE as a cause of the great flood.

Eventually, the twins went back to the poles and the earth stopped spinning. Most of the ice melted and Sótuknang created the third world. The people were ordered to live in oneness and to sing praises to the Creator. They emerged into Kuskurza (no meaning), the third world, and built "big cities, countries, a whole civilization." [23] This was the time of the ancient world, the time of the expansion of peoples around the globe. It was very hard in this world to remain spiritual, and more and more the Hopi focused on earthly things. A woman's wickedness is described; the people are now patriarchal. There was technology and war reminiscent of Atlantis' war with India:

> Some of them made a *pátuwvota* (shield made of hide) and with their creative power made it fly through the air. On this many of the people flew to a big city, attacked it, and returned so fast no one knew where they came from. Soon the people of many cities and countries were making *pátuwvotas* and flying on them to

attack one another. So corruption and war came to the Third World.... [24]

The Atlanteans had aircraft, which they used in making war. Perhaps "a shield made of hide" is the only way to describe a rocket or spaceship in the ancient Hopi language. A few people remained pure of heart and when Sótuknang decided to destroy this third world, he asked Spider Woman to place them inside hollow reeds, where they waited out the next destruction. The earth this time was destroyed by water, the 10,300 BCE great flood was also the last sinking of Atlantis.

> He loosed the waters upon the earth. Waves higher than mountains rolled in upon the land. Continents broke asunder and sank beneath the seas. And still the rains fell, the waves rolled in. [25]

The Hopi found themselves on top of a mountain, with no other dry land anywhere. They sent out birds that could not land, and finally built boats of the reeds that had sheltered them. They floated out to find the fourth world. Traveling long by boat and on foot, they finally reached Túwaqachi, World Complete, the present earth.

> It is not all beautiful and easy like the previous ones. It has height and depth, heat and cold, beauty and barrenness; it has everything for you to choose from. What you choose will determine if this time you can carry out the plan of Creation on it or whether it must in time be destroyed, too. Now you will separate and go different ways to claim all the earth for the Creator. Each group of you will follow your own star until it stops. There you will settle... [26]

In their travels to find the present world, the Hopi reached North America, crossing islands that had been the mountaintops of the third world and that sank behind them. These could have been the now underwater remains of either Mu or Atlantis. The Hopi story of the deluge sounds very much like that of a dozen other cultures, including the version in the Bible. Led by Spider Woman, the people reached the present land, but were not allowed to settle yet. They were sent on migrations to cross the continent four times, once in each direction, before they could return to what is now the stark Hopi land in Arizona. The abandoned Pueblos and great stone cities of the American southwest and South America are believed to have been stops on the migrations, as are the serpent mounds of the midwest. The Mayas of South America were a tribe of Hopis who did not complete their migrations. The Hopi creation and emergence story is perhaps the oldest record we have of life on earth.

The meaning of the fourth world, for the Hopi and for all of us, is choice. The cultures that could not live in oneness, "harming none" in peace and equality, were destroyed. We are asked to create a women's New Age of this world, an earth of peace, balance, respect and harmony. If we cannot, this world will end, whether by fire, ice, water or in some other way. Having relived the first three worlds in the development of this modern age, we are at the crisis point, the time of choosing. Hopi prophecies state that the end of the fourth world is near at hand.

José Argüelles writes about the end of the fourth world and what may be the most mysterious people in earth's story. The Classic Mayas of South America are people originally believed to have been a colony of Mu before its 50,000 BCE destruction, who merged later with survivors from Atlantis. Their stone cities are highly advanced enough to indicate Atlantean technology. They possessed an astronomical calendar that is the most complex and mathematically precise in the known world, and which is still not entirely understood today.

In his book, *The Mayan Factor* (Bear and Co., 1987), Argüelles' theory is that the 5,125-year Mayan calendar is a galactic setting out of the plan of creation, taking up where the Hopi story of the worlds and migrations ends. The current Mayan calendar cycle sets out the path of the world from 3113 BCE until 2012 CE, as a way of leading earth from the last days of the fourth world and into the fifth, the New Age. Earlier cycles described the passages through the four worlds, and if the fourth world began at the 10,300 BCE sinking of Atlantis, we are in the third 5,125-year Mayan cycle since the end of the third world.

Says Argüelles:

> Clearly, more important for the Classic Maya than territoriality and making war was the need to track the cycles of planet Earth by means of a unique mathematical system. The purpose of this elaborate record-keeping seems to have been the correlation of terrestrial and other planetary cycles within our solar system with the harmonic matrix of a master program. This matrix, encompassing the cyclical harmonics of the planets within our solar system, was *galactic,* as it represented a larger, more encompassing view than could be obtained from within our solar system. Unique by any known standard, this perspective implies that the Classic Maya were possessed with a distinct mission. Anyone with a mission also has a message — a fact which seems quite obvious, but all too often escapes the mind of materialistic archeologists....
>
> This mission, it seems, was to place Earth and its solar system in synchronization with a larger galactic community. That is the meaning of the dates and their accompanying heiroglyphs.[27]

Mayan art, like much ancient art, contains a number of illustrations of rockets and astronauts, fully recognizable as such today. These advanced people, who seemed to disappear by the tenth century, may have left the earth by spaceship. They may have been the people the Atlanteans were in contact with (from the Pleiades or Arcturus, Argüelles says). The abandoned Mayan cities are considered by the Hopi to be stops on their own migrations, and Frank

Waters believes that the Mayas are Hopi tribes that did not complete their wanderings (which would have returned them to Arizona). They could have been Atlanteans who left their cities to go elsewhere, in North or South America, back to Atlantis, or to other solar systems. The Classic Maya people and their highly complex calendar were perhaps a continuation of Atlantean technology, of galactic technology, and/or of the Hopi creation cycle.

If the calendar is, as Argüelles says, a synchronization of earth with other solar systems, the ending of this current cycle in 2012 may mean that earth is about to be recontacted, made again part of the plan of the Goddess universe. This could be a part of the choice of the new world. If earth was quarantined from contact with other systems because of our development away from oneness and "harming none," how we enter the New Age could be crucial to the recontact taking place. Atlantis lost galactic contact when its larger part was destroyed in 26,000 BCE, the end of the second world. Contact may have continued to the end of the third world when the Nefilim left Sumer, and possibly later, extending through the time of Babylonia, ancient Egypt and the Dogon. It may have been later still, with the seeming disappearance of the Classic Maya, which Argüelles puts at the tenth century.

One further culture needs to be looked at in this discussion of the third world, the world after Mu and Atlantis, and that is Crete. This matriarchal culture may have been a late colony of Atlantis, established at the third destruction. Goddess art herstorian Elinor W. Gadon says Crete was established by 7,000 BCE as a Goddess matriarchy. The patriarchal Mycenaeans overran it in the late fifteenth century (1500) BCE "when their civilization had been weakened by the devastations of earthquakes, floods and fires."[28] Crete's destruction may be compared to the 10,300 BCE ending of Atlantis, which it probably survived. Other earth changes happened at the time of Crete's ending, including the 1450 BCE destruction of Thera, a Minoan oracle center. Thera has been posited by some to be Atlantis itself, but the evidence refutes this.

Crete's symbols are similar to those of other cultures connected with Mu and Atlantis, and are the central symbols of Goddess matriarchy. The Cretan labyrinth is identical to the symbol used for emergence by the Hopi, and the labyrinth/spiral is also a womb symbol (or placenta?!). Crete's bull-dancing is believed to have come from Atlantis, and the bull (or more probably the horned cow) is also a Goddess symbol in Egypt, Africa, Greece and India.[29] Despite its connection today with bullfighting, there was no art in Crete that depicted the death of the bull. The Goddess Demeter is from Crete, as was the god Zeus, but though the culture was clearly matriarchal and Goddess-based, most of their Goddesses' names are today unknown. The Minoan Goddess is often shown holding serpents, a symbol of Mu and probably Atlantis' creation story, as well as of most other Goddess cultures. Deriving from Atlantis and existing until recorded time, Crete may be a last living picture of Atlantis and the third world, the time of the ending of matriarchy in known cultures.

Robert Temple, in his book on the Dogon, *The Sirius Mystery,* connects the mythologies of ancient Egypt, Sumer and Africa, and the placement of the eight Greek oracle centers, to Sirius the Dog Star.[30] Crete was a forerunner of Mycenaean Greece, whose culture in turn was germinal for civilization in the modern West. Theseus is said to have conquered Crete for the patriarchal Greeks, after going there as a tribute to the bull arenas and winning (or was it raping and abducting?) Ariadne, Crete's priestess/queen. Minoan Crete was a source of the pre-Hellenic Greek culture and religion, Greece before the patriarchy. Though the time of Crete's destruction is put later than the earth changes of 10,300 BCE Atlantis, a description is in keeping with the material of this chapter and the ending of the third world.

Barbara Hand Clow, describing her past life regressions as the priestess Aspasia of Crete, witnessed the earth changes of Aspasia's time. Her account in *Eye of the Centaur* (Llewellyn, 1986) of the oracle and the events is given here to make the idea of a changing earth real to women reading about it today. The changing of the earth and the world, as seen by Aspasia, connects the destructions of the three worlds to our own possible ending of the fourth. Her words are a conclusion to the past and an entrance into the present day.

> The first sensation I get is the sky seems to be getting really red, a strange red, almost blood-like quality to the sky. And then our hearing is almost destroyed by an incredible imploding blast! A few people fall, clutching their heads from the pain. What could it be?
>
> ...the earth is beginning to shake. There is a rumbling sound, a sound of rocks crunching underneath where we stand. Some fissures develop in the temple building. We feel very strange. The air thickens more. As a group, our consciousness centers. It is a mistake to be here; we know it, but there is nothing we can do about it now. We are just becoming fascinated and centered in the power of the Mother, as she seems to be writhing like a woman near orgasm. We hold our energies and watch, all aware that we are chosen to observe cosmic chaos. We are hypnotized....
>
> The temple is coming down. I can hear people screaming inside the temple. I am out in the back. People are clutching at my shoulders....The rock pitches me forward, and I fall out in front of myself, and then I'm on top of some people. And people are clutching. And then the rocks come down....I think what happens is I don't feel any pain because I've been hit right on the back of my head. It's like instant death. And I immediately get out of my body, instantly, because I want to see what the water is doing. And I'm curious about what the ocean is doing, what the rivers are doing, because it seems to me that if I'm going to understand at all what happened here, the tides and the ocean will tell me what happens....[31]

From out-of-body, Aspasia watches the destruction of Crete, knowing that she and the earth will be reborn once more. She has channeled the oracle before the earthquake:

Listen now and understand what I say. Soon a cataclysm will overwhelm you. The goddess religion is in power now. The goddess power will be blamed for this cataclysm, but no one on earth is responsible. It is simply an extraterrestrial cycle which manifests every 3,500 years. The patriarchy will assume power because the next culture will assume that safety exists in patriarchal order and control. And then the patriarchy will be blamed in a similar way when the cataclysm cycle repeats around 2,000 AD. Humans will only be free when they recognize natural cycles and let go of fear when they understand they will reincarnate anyway. [32]

Not so easy to do....

Prophecies say that we are now at the ending of the fourth world, and that the world to come after it will be an Age of Women, a New Age, a fifth world or fourth dimension. The prophecies also say that all is choice. The patriarchy of the fourth world can end in universal destruction, or it can end in peaceful change. We are the images of Goddess, whose thought power created the universe, and our ability to manifest and create will decide the world to come. We are already in the cycle of change, entering the new world. In our hands, in our knowledge of women's values and awareness of "harming none," we hold the changing earth.

Notes

1. Merlin Stone, *When God Was A Woman* (New York: Harcourt Brace Jovanovich, 1976), p. xxii.
2. Marion Webb-Former, *On Atlantis,* unpublished.
3. *Ibid.,* p. 4.
4. Lytle Robinson, *Edgar Cayce's Story of the Origin and Descent of Man,* p. 78.
5. *Ibid.,* p. 79.
6. Marion Webb-Former, *On Atlantis,* p. 7.
7. Zecharia Sitchin, *The 12th Planet,* p. 377.
8. *Ibid.,* p. 381.
9. *Ibid.,* p. 398.
10. *Ibid.*
11. *Ibid.,* p. 410.
12. Ruth Montgomery, *The World Before,* pp. 157-158.
13. Robert K.G. Temple, *The Sirius Mystery* (Rochester, VT: Destiny Books, 1976), pp. 20-21, 26.
14. *Ibid.,* p. 30.
15. *Ibid.,* pp. 31-32.
16. *Ibid.,* p. 32.
17. *Ibid.,* p. 215.
18. Frank Waters, *Book of the Hopi,* p. 15.
19. *Ibid.,* p. 16.
20. *Ibid.,* p. 19.
21. *Ibid.*
22. *Ibid.,* p. 20.
23. *Ibid.,* p. 22.

24. *Ibid.*, p. 23.
25. *Ibid.*
26. *Ibid.*, p. 27.
27. José Argüelles, *The Mayan Factor,* p. 50.
28. Elinor W. Gadon, *The Once and Future Goddess,* p. 103.
29. Buffy Johnson, *Lady of the Beasts: Ancient Images of the Goddess and Her Sacred Animals* (San Francisco: Harper and Row Publishers, 1988), pp. 294-295.
30. Robert K.G. Temple, *The Sirius Mystery,* p. 125.
31. Barbara Hand Clow, *Eye of the Centaur,* pp. 195-197.
32. *Ibid.*, p. 194.

The Present

The Fourth World

New Age Prophecies of World's End

There are times, sisters...
> when we are afraid that we will die
> and take the whole great humming dance of life
> with us
Something must change, we know that
But are we strong enough?
And will we be given time?

Starhawk[1]

The New Age has adopted the theory that the earth is about to go through a major, violent change. Not everyone is aware of the idea, but most who are either believe in it fully or disbelieve it totally. Some want to deny it, not look at it or talk about it, but are made uncomfortable by it anyway. Others see the earth changes as something coming, but not now, though many prophecies make it a feature of the next ten or twenty years. Some see it as something outside of them, something that will affect others but not themselves. Or maybe it just won't happen, it hasn't happened yet. Maybe it isn't real.

There are people who see disaster as the only way for positive change to manifest in the world, and believe that change can only happen traumatically. The earth changes will make a new societal system possible and end war. Nothing else would be drastic enough to do it. They believe that reduced population and an end to overcrowding would be a good thing, but that mass devastation is the only way it can happen. The old, ongoing abuses to this planet and its people will end, and everyone will be fed and free, but only by devastating means. Where people used to fear a nuclear war, they focus on earth changes now, and talk of "safe areas" to go to to wait it out, hoping destruction will pass them by. The Appalachians and Arizona are safe, but California, New York City and Florida are not. The midwest may be safe, but not near the Mississippi or Great Lakes. Canada is safe. And what about places near nuclear power plants? Few who discuss it are relocating, though.

There are people who envision the meaning of their world breaking down.

They see themselves as explorers and pioneers, but fear the hardships of frontier living. They see themselves as nevertheless able to be self-reliant, free of government interference and taxes, and growing their own food. The government will take less away from them then, but it will take a disaster to accomplish that. Some people now store food, while others feel more secure and don't, and some worry that they won't know what to do. Some read every book on earth change theory they can find, consult every channeler and read science fiction, seeing themselves in all the disaster scenes and roles.

Fundamentalist Christians, taking their cue from a confused reading of Revelations in the Bible, believe that earth changes means the world will come to an end. They see themselves as being "translated to heaven" on judgment day, and gear their actions in the world to hastening its final ending. The "end of the world" is desirable to them, for no matter how much destruction, they alone will be saved. Each patriarch believes that he is one of the chosen, but his neighbor is not. The Jehovah's Witnesses believe that only 144,000 will be saved, and each expects to be one of them — there are two million Jehovah's Witnesses, by their own count. Some New Agers make metaphysics into another brand of Christianity and implore their god for mercy, or implore the earth herself. They ask that there won't be change in the world, or that they themselves can avoid it.

All of these attitudes are passive and simplistic. The popular New Age believes that the world will end, and that individuals have no part in the changes and can do nothing about them. They wait for the day and hope it won't come, or that they'll be somewhere safe if it happens. They can imagine others dying and others' cities or homes being destroyed, but not their own or themselves. Few of them have thought long enough to see the earth changes as *change,* as a process of development and growth, or as a new beginning, another chance. Few have considered that the world may not end at all, or thought about what comes after deep inner growing. Few have any idea of what a New Age could be, though they talk of "paradise" and "no more war." With their passive stance, few New Agers work toward achieving the things needed on the planet, righting the wrongs so that violence ends and destruction doesn't have to happen. The New Age public has been raised on television — they sit and watch, having no idea that there's anything they can do.

New Agers who accept the idea of earth changes also see no way of changing the prophecies about them, and they accept the predictions as inevitable. They hope the psychic who talks about destruction is wrong, but disbelieve her if she talks of choice and creating positive options. They put negative predictions as far away into the future as possible to make them safe enough to think about. There is little attempt to use prophecy to define what needs changing and then to work on those things and change them peacefully. Most have no idea that even earth changes are subject to the Goddess' law of cause and effect. What is predicted will or will not happen, with no thought of why or why not. If a prediction fails, it's because the psychic was wrong, not because the situation changed, or because the situation *was* changed by thoughtful action.

The women's movement and women's New Age does not wait helplessly for disasters, or believe that the world will end. There is too much to do to "save the planet," "end global warfare" and effect peaceful change in the world. And since the beginnings of feminism, women have proven time and time again that they can work miracles. Instead of waiting for destruction and hoping it won't happen, women whose Goddess thought power made the universe set about to change what needs changing and clean up the earth. They work against patriarchal aggression and greed, against racism and poverty and misogyny. They work at reducing pollution, and at boycotting corporations that are stubborn offenders. They work at raising human consciousness about what's needed, especially the need for "harming none" and caring about the planet and each other.

The most valuable prophecies for women are the ones of dooms that don't happen. Every negative foreknowing is given with the idea that this is what will be only if we don't change it. By knowing what's expected to come, women are given information on what needs changing and they do their best. The number of people in the 1980s working for world peace was so great that channelers now feel that the long-predicted global war will not happen. Such concentrations of women's thought power can make positive changes in other areas, in all the things that need "fixing." By visualizing what a positive New Age means and working toward it, other destructions can be averted. No prophecy is written in stone, and the predictions themselves are changing. They are guideposts for the "change" in "earth changes."

The subject of this section is the predictions channelers have made in the recent past regarding potential earth changes to come. I avoid Christianity here, to quote instead known psychics. Most of the material in this chapter is from books rather than direct sources, with recent channelers later. The predictions given are the ideas that have shaped the New Age conception of earth changes, and as always I use women's writings wherever possible. Be aware that more recent channelings and information, readings after 1987 generally, paint a much gentler and more complex picture than the material given earlier. Be aware also, that no prophecy is bound to come true just because it is made. Every day that we live on the planet new input causes changes in any information. We live in a flowing river. Be also aware that women's power of thought, our Goddess-within ability to create our lives and world, can change most negative predictions for the better.

I do not see the concept of earth changes as a prediction of the end of the world. Rather, I see it as a major energy shift and transition for planet earth, a shaking off of the wrongs of patriarchy and an instituting of women's "harm none" values for a Goddess New Age. I see the earth changes as the planet growing up, maturing, letting go of its adolescent male abuse of life, and rejoining the league of intelligent Be-ings in the universe. The earth changes for me are a new beginning, and another chance for women's culture to exist on a Gaea that can appreciate and live well through them. If the birth of this new world is a slow process, so are most births from the Mother; the results when born are worth the waiting and the effort. Visualize the Ten of Discs in the Motherpeace tarot deck, a new birth after long labor, made in a circle of loving women. I do not see the earth destroyed in the coming changes; I see her purified, changed and renewed, and a women's civilization also renewed and made whole again. This is what women are working toward.

New Age earth change prophecy takes the form of a series of steps in the process of ending the old ways and instituting new ones. The 1980 book *Rolling Thunder* by J.R. Jochmans (Sun Books) gathers together a number of earth change predictions from a variety of sources, and gives the following general outline of things to come. Remember that this is "classic New Age" earth change theory, and also that predictions have shifted considerably since 1980. The severity here is typical of earlier information, where cataclysm and catastrophe were emphasized, instead of today's (and women's) emphasis on free will and choice. I cannot depict and describe these disasters without also reminding women that what they create will happen. Therefore, in visualizing, see how the disasters are prevented, rather than adding energy to them.

Classic New Age theory presents a series of seven steps on the path to a new world. These are:

1. Period of Disintegration
2. Period of Catastrophe
3. Period of Restoration
4. Period of Domination
5. Second period of Disintegration
6. Period of Termination
7. Period of Re-Creation[2]

Timing for this process varies, but many psychics date a World War as happening in the 1990s, and a pole shift occurring anywhere from 1996 to 2050. Some psychics say the timeframe is speeding up.

The *Period of Disintegration* predicts a time of runaway inflation and economic depression, leading to economic collapse and the "collapse of interdependent world economies." There will be the beginnings of totalitarianism in governments that claim to be democracies, along with the collapse of old ideologies, both democratic and communist. Nations will be polarized into two blocs, that of the overdeveloped West and Russia versus the Far and Middle East with other nonwhite and newly developing nations. There will be increasing corruption in the West in the forms of overblown materialism, sensationalism, greed and nihilism. There will be radical change in the role of churches and patriarchal religions, with increasing government collusion in church issues of "morality" and repressive religious values in a decadent atmosphere. To balance this: an increase in New Age spiritualities and growth in the population that practices them.

A lot of this looks like the daily news. Inflation is no surprise to anyone living in the West, from food and clothing, to housing and utility prices, to anything else. Each paycheck, when there is one, goes less far. And while governments routinely deny that depression or recession exists, the number of homeless in America was large enough that the 1990 Census sent representatives into the shelters to count them. (The figures are not forthcoming.) Most of those on welfare in the West are mothers with small children, and there are all too many women and children in poverty, especially people of color. There is a definite increase in conservatism and legal repression in restricting reproductive freedom to blocking the release of promising AIDS medications, in the erosion of women's and gay rights to punitive tax increases that consistently favor the rich and give the poor less to live on. America may be freer than some countries, but it is not free, and is becoming less and less a democracy. This is also increasingly true of other "developed" Western nations.

The weakening of communism and apartheid have been exciting developments since 1989. The once-communist nations of Eastern Europe are instituting a democracy that will not be American but a form of their own. South African apartheid is coming apart piece by piece. There are increasing liaisons between nations of the West and the Soviet Union, and increasing difficulties with Eastern nations. Brutal repressions of the pro-democracy movement in China and fanatical terrorism in the Middle East have created an ever-widening division between East and West. One of the more disturbing factors in the current situation is the rise of religious fundamentalism in a variety of right-wing religions, and their entrance into governments from Iran to the United States. America is certainly caught up in such religious "morality" campaigns and in political repressions prohibited by the Constitution. Reproductive control is only one issue in this right-wing campaign to erode the rights of women and put Americans under church domination. Religious fundamentalists in other countries, from Iran to Israel, are causing worldwide repression, terrorism and intrigue.

An increase in New Age spiritualities is also evident in the West, from the hype of prosperity consciousness workshops to the more serious self-empowerment and spirituality as political change movements. There has been a significant increase of the New Age in all facets of daily life, from the number of metaphysical bookstores and events, to the awareness of New Age thinking in the mainstream. Women's Spirituality and wicce have grown exponentially in the women's movement, as have Eastern religions, wholistic healing, channeling, astrology, and studies of Native American religion.

A *Period of Catastrophe* is the second phase of the earth change scenario. This is defined as ecological disasters on land and sea, with worldwide weather aberrations, crop failures and food shortages. A World War between the East and West is described, using limited nuclear and/or bacteriological warfare, and leading to a complete collapse of Western civilization. Jochmans paints a picture of chaos and revolution, the total decline of culture and technology, and total government breakdown. There is a period of local governmental rule and survival living. In an atmosphere of pessimism, religious revival becomes a mass movement, with increasing talk of a messiah. Right-wing religious leaders will become more aggressive, and religious power totalitarian. To balance all this will be the appearance of a few true teachers, spirituality leaders who bring sanity and comfort but have no interest in taking power.

As I write this in 1990, weather aberrations are an increasing news item, with lower than average rainfall producing crop reductions in some places and floods putting many out of their homes in others. The summer of 1988 brought intense heat and crop failures in the midwest, and Ethiopia and Bangladesh have experienced alternating floods, famines and dis-eases for several years now. Ecological disasters are happening with increasing severity, from a meltdown at the Chernobyl nuclear power plant in 1986, to the 1989 Alaskan oil spill. Something appears on the news almost every day about an oil spill, garbage barge, animal or human epidemic, or earthquakes or tornados. While most of today's channelers believe that the threat of nuclear war will not manifest, the repressions in China and terrorism in the Middle East keep everyone uncomfortable. We are near chaos and revolution in a number of places in the world daily, and more is predicted to come. The atmosphere of pessimism and religious revival is here, and even with the debunking of the televangelists it has yet to decline. The right-wing continues to be aggressive, and we shall see how far the Supreme Court and popular opinion lets them go. There is the balancing factor of a few spiritual, unselfish teachers with no use for power or control; these individuals have always been among us for those who can hear them.

The *Period of Restoration* is next, going beyond what can be assessed with today's information. In this period there is economic recovery, with new centers of economic power and a gradual control of world economy by the churches. Governments are re-established which are weakly totalitarian and run by religious leaders. The dominating worldwide atmosphere is patriarchal, right-wing and religious, with the beginnings of a "one-world religio-political monopolist government" that contains the features of "ideational, pseudo-mystic, monastic, mass-hypnotic, irrational salvationism."[3] There will be collapses of current patriarchal religions and an ultra-patriarchal global church will emerge to pressure Eastern and other religions to disband or compromise and be absorbed into fundamentalism. A pseudo-messiah will be media-advertised and pro-moted by this world church. The balancing factor will be that New Age spiritualities will go underground to continue and preserve their message and ancient teachings. They will warn that the messiah is a fraud.

In the *Period of Domination,* there will be a stable world economy and stable totalitarian governments, all under church domination. Global peace will exist, but with repression and lack of personal freedoms as the price. Control of the masses and individuals will be the means of maintaining church/state power. Any opponent to the system will be labeled "heretic" or "anti-Christian," and there will be forced conversion of all religions and philosophies to the global church in the name of world peace. This will create one universal religion and government with complete totalitarian control, handing out death penalties to nonconformists. Power will be centered in an international messiah figure. To balance this: true spirituality leaders will continue to exist and teach, despite persecution. Many people will adhere to their own religions despite the power of the global church, and secret study of ancient religions will continue.

The *Second Period of Disintegration* will focus on the increasing greed, rapaciousness, and materialism of the new church, as a strong wave of ecological disaster puts the world in an economic depression. Finally there is a revolt of world leadership against the global church, and an end to its power and control. There is a realization of the decadence, emptiness, hypocrisy, spiritual wasteland, and universal despair of the church system. The "Babylon Civilization"—right-wing, church-dominated totalitarianism at the expense of individual free-dom — ends. In balance: spirituality prevails over the abuses of the era. Spiritual leaders warn of a coming earth cleansing and lead their students to safe areas.

The *Period of Termination* is a time of global chaos and general disorder, marked by attempts of the global church and its pseudo messiah to regain control. There will be conflict among factions for power in the world, and preparations for a world war:

> Spiritual forces of Light and Darkness are seen fighting in the heavens. The earth shifts on its axis. Mass destruction. Warring forces on the earth are annihilated. Triumph of the Light.[4]

There is no balancing force or factor.

In the last period, the *Period of Re-Creation,* the true spiritual teachers and their students will emerge from underground hiding to rebuild. This is the classic vision of a New Age, vague though it is to specific forms and meaning:

> Rebuilding is begun, in harmony with the earth. Founding of the Aquarian Spirit-Age. Present forms of religious and political power are abandoned and forgotten. The New Civilization of Light emerges: Spiritual, intuitive, super-sensory, transcendental. The appearance of True World Saviors. The brotherhood of man

[sic], and the communion with the Godhead, and Spiritual progression continues unlimited....[5]

This is male as usual, a "brotherhood of man," and communion with the Godhead." The New Age has no other vision of a different form of world than the patriarchal more-of-the-same. Women can create something better, specific rather than vague, "harming none," and based on respect and freedom for all. What this New Age might be is discussed in the last section of this book, The Future.

The seven steps are a compilation of classic New Age theory, rather than information from any one source. Most of the psychics and theorists are women, as are most participants in New Age spiritualities, but without a Goddess or feminist awareness. Ruth Montgomery, in her 1979 book, *Strangers Among Us,* draws a picture that fits with the compilation above. In 1982, in her book *Threshold to Tomorrow,* she looks at the predictions again. The early dates are important here, as much of the severity of earth change prophecy is lessened after the 1987 Harmonic Convergence. These later predictions are discussed in the next chapters.

Ruth Montgomery's view of the change from the Age of Pisces to the Age of Aquarius includes World War III unless "herculean efforts" are made by enough peaceful people to prevent it.[6] She sees it as possible to prevent, but feels that limited nuclear war seems likely. It will be entered into by leaders seeking a way out of domestic problems, as a distraction from the ills and economies of society at home. Ethiopia is the starting place of a smaller war that expands into a global one. The war or its potential, will not last beyond the early 1990s.

The potential is definitely there. Famine has existed in Ethiopia for several years, with relief efforts blocked by the Tigre rebels and the country's civil war. The entrance of super-powers, in the name of opening the roads for humanitarian aid, could conceivably escalate. At this time it does not look likely, however. The potential is definitely there, however in the Middle East, with Iraq's invasion of Kuwait and the rapacious war with the US that devastated both countries in 1991.

Montgomery's analysis urges people to begin planning and living in survival communities, moving toward self-reliance. She feels these communities will spring up with ever greater frequency in coming years, as more and more people flee the decay of the cities for rural life. Local food production will be a hedge against famine and inflation, as well as training grounds for cooperative living and New Age thought. Enough Americans have the pioneer urge for frontier living to make this alternative lifestyle seductive.

From the perspective of the women's movement, this back-to-the-land idea is nothing new — it has been developing steadily since the 1970s. Male and female communes were a feature of the 1960s back-to-the-land movement; some of these still exist and thrive. As life in the cities becomes more dangerous and violent, less life-affirming and more polluted, returning to the earth is a healthy option for some people. New Age retreats, both mixed and women-only, are also developing, and while most of these are not yet residential communes, some of them are. Much feminist utopian literature involves a return to rural living and farming, and the United States and Europe have a strong history of intentional communities and experiments over the years. This type of living is not for just anyone, but for those with physical strength who appreciate living simply, cooperatively and with a minimum of technology, it is definitely a viable alternative. There are still places, even in the overpriced West, where land can be obtained affordably.

In the late 1990s, says Ruth Montgomery, increasing signs will be seen of a coming pole shift. There will be scientific evidence of wobble in the earth's orbit around the sun, and an increased interest and discussion in finding "safe areas." Increasingly violent weather will be a feature of this coming shift, with record-breaking snowfalls, cold, gale-force winds and increasing general humidity. Summers will shorten in the northern hemisphere, and there will be changes in atmospheric pressure. There will be volcanic activity in North America, the Caribbean and the Mediterranean, and increasingly frequent and severe earthquakes globally.[7] The weather activity will also be caused by pollution and damage to the ozone layer, with the health of many people adversely affected by both pollution and inclement weather.

Severe weather has been an increasing feature of the daily news for the past several years now, including high winds, flooding, droughts, and extremes of heat or cold. The environment, damage to the ozone layer, acid rain, the cutting down of oxygen-producing forests, and the effects of pollutants and toxins are also entering general awareness. Healers have warned of the effects of polluted air, soil and water for many years. Rates of cancer are increasing, including doubled rates of skin cancer from the lack of ozone-layer protection, and cancers in the children of migrant-worker mothers exposed to pesticides. The degraded environment is also hosting a new class of dis-eases, immune system dysfunction dis-eases, including allergies, AIDS, multiple sclerosis, lupus, arthritis, and chronic fatigue syndrome. Too many people are being affected, significantly women, gays, the poor and nonwhite, with no solutions forthcoming.

There is little movement to clean up the earth or restrict and penalize the polluters. These things were just beginning to be known in 1982, when *Threshold to Tomorrow* was written. The number of damaging and life-taking earthquakes has seemed to increase around the world, and there have been volcanic eruptions in North America of Mount St. Helens, Kileauea and the Redout Volcano in Alaska.

"Safe areas" for Ruth Montgomery include places away from the coasts and Great Lakes, with mountain ranges existing between the safe area and the ocean. Central North America is listed as a secure place, and most of Canada, as well as the American Northwest. Indiana, Oklahoma and Arkansas are listed as safe, and Illinois away from the Mississippi River. Europe, particularly on the coastlines, will be less safe, with flooding, high winds and tidal waves predicted. England and Holland, says Montgomery, will be totally submerged, along with other coastal areas. Land areas in the southern hemisphere and in Scandinavia will expand, with New Zealand and Australia increasing in size as well. Some Pacific islands, including Hawaii, are predicted to disappear. Canada and Russia will develop a warmer climate; so will parts of the United States, but parts of the United States will also get colder. South America will have earthquakes and land changes, and the Antarctic icecap will melt again. [8] Parts of California and New York City will sink into the ocean, and the Mississippi River will enlarge as the Great Lakes shift to empty into it. The land around the Mississippi will flood.

Montgomery says that souls who existed on Atlantis have been reincarnating in great numbers on the earth, especially in America, for the past fifty years. A number of new incarnations will be formerly from Mu, and these peaceful people will be the root of a New Age culture on a changed earth. More and more of these are being born since 1978 and will continue to be. There will be more awareness of reincarnation and connection with the spirit world, and more matriarchal/women's values of cooperation, harming none and peace. There will be communication and contact with other planets. [9] She mentions an anti-Christ that will be overcome, and the coming of a messiah. In her earlier *Strangers Among Us,* she paints an earth-change scenario very much like that of the Jochmans list, with few changes in her later book.

Another psychic and channeler, Page Bryant, published her earth change predictions in *The Earth Changes Survival Handbook* (Sun Books) of 1983. In it she describes the coming changes as part of the planet's process of growth toward self-realization and as an initiation process. This process began in 1976 and will conclude around the year 3000, with the most major activity occurring between 1986 and 2050. [10] She describes earthquakes, storms and volcanic eruptions as indicative of the beginnings of the earth changes, and judges the predominant factor to be climate. She predicted India to experience an increasing population and famine problem in the 1980s and 1990s, with droughts and food shortages. The Soviet Union will be one of the hardest hit areas for food shortages and will attempt to befriend other nations to obtain what it needs. She predicted droughts that are short-term in the United States in the 1980s and 1990s, but severe enough to cause economic hardship and food shortages. Canada will experience water shortages, while China will have serious flooding; flooding will occur in the United States in Florida and along the Mississippi

Delta as well.

Page Bryant's guide Albion describes volcanic and earthquake activity in Japan, with catastrophic destruction there by the year 2025. Like Japan, Samoa will sink beneath the sea. There will be ice melts in Greenland, and a pole shift will change it to a subtropical climate by the year 3000. There will be an increase in lightning storms, with a form of lightning that is violent and highly dangerous. By the year 2000, city living will become mostly undesirable, and more and more people will join rural communes. [11]

Some of Page Bryant's predictions have been proven as in the one of famines in India and Russia. The Soviet Union is indeed turning to old enemies for solutions, and changing its government drastically to alleviate the shortages, long lines and paperwork that the people have had enough of. Bryant states that much of the Soviet Union's political actions are based on its food needs, which also is proving true in the 1990s. Short-term but economically significant droughts in the United States grainbelt have also begun, notably the 1988 summer of excessive heat and low rainfall that destroyed a large part of the American crop and caused high prices and great hardship for farmers. Spring of 1990 has brought severe flooding to Georgia, Florida, Arkansas, Oklahoma, Texas and the Mississippi Delta, along with tornado damage in Kentucky and Indiana and less severe flooding in Ohio. A cyclone in 1991 in Bangladesh left 138,000 dead.

Earthquake activity is also on the increase with the Iranian earthquake leaving 40,000 dead in June, 1990 followed by another in 1991. An earthquake in Armenia in December, 1988, left 25,000 dead, and the October, 1989, San Francisco earthquake killed 62 and caused millions of dollars in property damage. There have been earthquakes in China, a seaquake in Japan, and quakes in New Zealand, the Sudan, Peru, the Philippines, Roumania and Mexico, all higher than 6.5 on the Richter scale, and all in a period of less than a year (1989-1990). The Redout Volcano in Alaska has become active, and there has been new activity at Mount St. Helens in Washington State, as well as Kileauea in Hawaii (December, 1989–May, 1990). Many people who live in inner cities now would like to leave, and what is known as "white flight" to suburbia is a fact, with wealthier white families fleeing and abandoning the inner cities to decay. There has been an increase in rural living, but not a significant one, and in fact many farmers have been forced from their lands by agribusiness takeovers. Bryant says:

Whether the Earth changes are a reality or the prophecies of earthquake and volcanic disasters ever occur, only time will tell. It seems hardly coincidental that so many seers from the past ages up to the present would tell almost the same tale. The evidence is mounting regarding the "tell-tale" signs of the coming chaos....scientific and non-scientific, geological and historical records of both planet and man tell us that this has happened before and chances are it will happen again. [12]

She describes "Man himself" as the "biggest disaster of all," outlining the abuses of Western technology, greed and war. Like the misuse of the Firestone in Atlantis, men now have the technology to destroy the earth many times over with the push of a button. She cites accompanying problems of nuclear and toxic wastes, and pollutants from industry, the number of species becoming extinct, and the cutting down of the Amazon rainforest. "Man's" future lies in his own actions — and in women's ability to bring balance and healing to the devastation men have made on earth since patriarchy began.

Bryant emphasizes self-reliant alternatives as a must for dealing with the coming shortages and ecological disasters, a moving away from industrialization at least on an individual or communal basis. She talks of close-to-the-earth living and of learning to farm organically, as well as developing a variety of manual skills from composting and first aid to constructing buildings and canning food. Some of her instructions sound like the fallout shelter movement of the 1950s. The lists include dried foods and the means to store them, survival tools, first-aid supplies, clothing, seeds and equipment. She specifies as "safe areas" Arizona, New Mexico and Colorado, Washington State, Oklahoma, most areas in the Atlantic time zone, and western Michigan. Until the year 2000, Kentucky, Virginia, West Virginia, Tennessee, Georgia and the Carolinas will also be "safe," but not after that time.

Earlyne Chaney is a California psychic and mystic who channeled *Revelations of Things to Come* (Astara, Inc.) in 1952, and reissued it with updates in 1982. Her work is Christian-based, focusing on the worldwide appearances of Mary as the Earth Mother. She sets her hope for a New Age on women's empowerment and awakening. Chaney's vision of the earth is that of the planet as a living Be-ing, something only now being popularly restated by James Lovelock in his "Gaia Hypothesis." Says Chaney:

> It must be remembered that our planet Earth is an entity, a great cosmic entity — a Mother Deva which has been almost destroyed by the lifestream of her inhabitants. Earth has been slowly poisoned and polluted by the greed, the selfishness, and the ignorance of the children to whom she has given birth, to whom she has offered her molecules of matter to create forms. [13]

She sees the earth changes as a purification and initiation of the planet, "whatever is necessary to purify her own being, her own etheric bloodstream, her own consciousness." [14] (The "she" pronouns are Chaney's.)

According to this woman psychic and mystic, these changes can come about gradually or cataclysmically, and are decisions of human choice. Through thought power and the raising of awareness, they can be less severe, but cannot be stopped. She lists the same upheavals of nature that other early psychics of the New Age depict, including World War III and a pole shift that may be gradual. The happenings are not random, but are carefully timed and planned to destroy/purify some irreclaimable areas for further renewal, and to spare others that are important for earth's rebirth or are ecologically pure. The cleansing can happen more gently with human help and consciousness. The earth will continue to support life, and continue to provide food for its inhabitants, whether the changes come drastically or not.

Beginning in 1950, both the planet and people have entered into this purification. The changes happening in the earth are reflected in the changes happening in individuals. Women's emotional purification is as intense as the planet's geophysical one and as painful. Some Be-ings, those not ready to make the emotional and awareness-level changes, or those who find it too hard to go through at this time, may make the conscious or nonconscious choice to leave the planet. The many deaths occurring, whether by cancer and AIDS or by earthquakes and tornados, should be seen in this light — as a choice. Death is a transition to a new state; in itself it is painless, and the soul is immortal. Remember the Goddess teachings of reincarnation and karmic theory. Any soul incarnated at this time has chosen to be here for the experience. Bodies die, but souls do not.

Only those souls with the "harm none" awareness to create a New Age will be allowed to reincarnate again on this planet. Those who cause harm and pain in the world, who do wrong and violence, will no longer be able to be here. They will not reincarnate on earth. Chaney attributes the tremendous increase in world population since 1930 to the number of souls wishing for a last chance to increase their embodied awareness before "the closing of the door":

> Until then, there will thus evolve two worlds, co-existing. One world will be engulfed in waves of darkness — increased crimes, murders, rapes, muggings, robbings....The other world will advance, like a vast army, toward a golden tomorrow....

> At a certain period there occurs what is called the closing of the Cosmic Door. This means laggard souls can no longer come into birth in this lifestream. It means advanced souls will begin to take on forms of matter. [15]

This closing of the door she believes to be the meaning of the biblical judgment day. The emotional purification and cleansing of individuals is partly a choice to accept or reject "the light" and partly a cleansing of karmic debts, which must be cleared for an individual soul to be able to reincarnate in the after-earth changes new world.

The emergence of the feminine, women's values of peace, love, healing and "harming none," had its modern beginnings for Chaney in 1846 with the first of the appearances of Mary, whom she also calls the Great Mother. These appearances occurred on sites known to pre-Christians as places sacred to Persephone or to other aspects of the Goddess. The first ones occurred at LaSalette in France, Fatima in Portugal, Garabandal in Spain, and San Damiano in Italy. Lourdes is perhaps the most famous, and the most recent of such visitations of the Lady have been in Medjugore, Czechoslovakia, in the last few years.

> The Lady's appearances marked the beginning of a forward thrust on the part of women to occupy a significant space in unfolding world events, albeit there may be no awareness on the part of the feminine consciousness that the Virgin was — or is — in any way connected with or responsible for the social transformation that will begin to gather momentum in the early 1970s. [16]

This was written in 1952, and can only refer to the feminist and Women's Spirituality movements. She attributes these movements to the entrance and awakening of the Earth Mother/Mary into consciousness again, particularly into the consciousness of women. Chaney sees women's growing empowerment, the rise of the women's (and unforeseen Women's Spirituality/Goddess) movement, as the key to world peace and to leading earth and its inhabitants through the coming time of purification and earth changes:

> As the time-compendium swings into the closing decades of the century, woman will have won her rights as an independent being with the power to quietly influence the race toward light, love and the liberated soul, capable of leadership and responsibility. And man will give way to recognizing his innate need for such a transformation and how much more purposeful is the world overview with woman standing as his equal. [17]

I believe Earlyne Chaney to have touched upon some key issues, and done so very early in New Age literature. Her prediction of the women's movement and the reawakening of the Earth Mother/Goddess needs no comment; it has happened and is happening, as clearly as she stated it. The appearances on earth of Mary, beginning in 1846 and ongoing to today, have presented something highly compelling for women and spirituality. Later in this book is the channeling of Laurel Steinhice, who is one of increasing numbers of psychics touching upon Earth Mother energy. Earlyne Chaney is doing something of the same thing here, and connecting the appearances of the Goddess as Mary to the concept of a living planet, something Women's Spirituality is based upon.

Another key issue here is the idea of women's internal purification as happening in concordance with the earth's. Since "You are Goddess" is a focal point of Women's Spirituality, the idea of individuals going through the same earth changes as the planet we live upon makes sense. It also explains the deep pain and inner transformations women have been going through for the past few years, changes emotional and spiritual. It is the concept of the closing of the door to incarnation, separating those who "harm none" from those who cause harm, that is resulting in the current increase in violence and negativity, being felt so tremendously today. That such negativity and negative Be-ings will not be permitted to return is a true relief, offering hope for both the New Age and for this one. The settling of karmic debts and the releasing of old pain is a part of women's transformations and is positive as well.

The awareness that many will die, leave the planet in the earth changes, is also an awareness that is painful but has an explanation within Chaney's well thought-out work. Remember that souls are immortal, and that death is a transition. Death is a moving-on to another plane, other work in other dimensions, planets or worlds. On the soul level, these deaths are made by choice. It is part of women's work today, healers' work, to help emotionally those passing over and to ease their physical suffering as much as possible. The acceptance of death as a transition and the idea of reincarnation makes the death process into something it can never be under patriarchal religions. As embodied souls, we were not meant to lose this awareness that patriarchy has taken from us.

Women are very much involved in changing this lack of awareness of death, as they are in every other aspect of the earth change consciousness. We are the people whose awareness is changing first, certainly influenced by Mary/the Earth Mother/Goddess. And by that awareness and creative thought power, it is women who lead the earth through the purification and into the New Age. This perspective gives some of the distress of the period — AIDS, cancer, violence and natural disasters — both meaning and comfort. Earlyne Chaney is one of the earliest of the classic New Age earth change writers. She usually offers the most sense and the most reason, as well as a Goddess awareness. Her analysis is far different from the patriarchal bias of most early earth change theory, although her predictions are much the same.

Notes

1. Starhawk, *Truth or Dare: Encounters with Power, Authority and Mystery* (San Francisco: Harper and Row Publishers, 1987), pp. 1-2.
2. J.R. Jochmans, *Rolling Thunder, The Coming Earth Changes* (Santa Fe: Sun Books, 1980), pp. 164-167.
3. *Ibid.*, p. 165.
4. *Ibid.*, p. 167.
5. *Ibid.*
6. Ruth Montgomery, *Threshold to Tomorrow* (New York: Fawcett Books, 1982), pp. 196-197.
7. *Ibid.*, p. 198.
8. *Ibid.*, pp. 199-200.
9. *Ibid.*, pp. 206-207.
10. Page Bryant, *The Earth Changes Survival Handbook* (Santa Fe: Sun Books, 1983), p. 280.
11. *Ibid.*, pp. 281-183.
12. *Ibid.*, p. 285.
13. Earlyne Chaney, *Revelations of Things to Come* (Upland, CA: Astara, Inc., 1982), p. 23.
14. *Ibid.*
15. *Ibid.*, p. 26.
16. *Ibid.*, p. 47.
17. *Ibid.*, p. 48.

Recent Women's Prophecies

1984-1987

Most feminine leaders have not yet become aware that the feminine forward thrust seemed to have its birthing when the great Lady of Light first began making her appearances and prophecies in various locations around the globe. Could it be that she — the foremost of all feminine forces — is the underlying influence behind the scenes of woman's release into New Age reality?

Earlyne Chaney[1]

The idea of the Goddess — Lady of Light in all her many names and aspects — as the underlying force behind the New Age, is a central idea of this book. The Goddess/Earth Mother exists as both the planet and those who live upon her. Whatever happens to the earth happens to women, and it is women who will create the Age of Aquarius, the Age of Women, the thousand years of peace and freedom that are the predicted outcome of the earth changes. It is also women whose "harm none" awareness and ability to teach it will determine just what form the earth changes take — whether those changes are to be gradual or sudden, cataclysmic or orderly, involving great loss of life or less. The Goddess teachings of "harm none," "you are Goddess," and women's consequence is crucial here. Free will and free choice were two of the central things made part of Be-ing at the time of souls' embodiment on the planet. How we use that free will now is the determining factor of our planet as a planet and as women.

The material of the last chapter presented classic New Age earth change theory, mostly by women, but also mostly lacking a feminist or Goddess consciousness. The predictions of this chapter are more recent, from about 1984-1987. They are marked by a far greater awareness of free will, of positive change, and of the earth as a living Be-ing/Goddess. They speak to modern life and what's apparent daily all around us, as much as to a far off destruction followed by a utopian someday. There is change in the nature of the predictions, but not as much change as after 1987 in readings by the most current channelers. The later the predictions, the more earth/Goddess awareness and the more focus on reasons for events, what can be done to help, and what is women's role.

Meredith Lady Young, a channeler whose guide's name is Mentor, writes of earth changes in her 1984 book *Agartha* (Stillpoint Publishing). She describes the ionosphere as earth's aura and attributes "negativity in the form of individual thought patterns magnified many billions of times"[2] as the precipitating factor for a planetary change and cleansing. Destruction of the ionosphere (and ozone layer), the protective envelope surrounding the planet and acting as a filter, would mean the end of all life on earth. The earth changes are the means for preventing this destruction by freeing the planet's aura of negativity.

> The thinning of the ionosphere...represents the given allotment of time for a developing world to shift from a self-orientation to a group-orientation, from pure analysis to feeling, from blindness [*sic*] to awareness. It is like a yoyo attached to a cord. At first there seems an unlimited amount of cord, but as the yoyo descends, one quite suddenly feels tension on the cord as it begins to run out....
>
> If Earth is the yoyo and is fast approaching the end of its cord, who is there that is sensitive to the tension on the cord?...Those with realized personal power will speak. Greater and greater numbers of people will feel the tension on the cord and will, through shared positive energy, allow the Earth to stall only temporarily before it begins its upward path....[3]

If despite being given every opportunity, this upswing does not happen; if humanity cannot switch from selfishness to oneness, the earth will not survive. The shifting of the earth changes is the shifting of human awareness to assure its survival. The means of earth shifting is in the movement of the tectonic plates, twenty of which make up the earth's land surface in geologic terms. The movement of the plates has a physical explanation, but the plates themselves are energy forces that mirror the energy of the continents above them in all their histories. This energy pattern is comprised of collective consciousness:

> Given that energy patterns once created continue to exist within the Earth's framework until transmuted and allowed to return to the universal stream of awareness, then each thought lives as a "real thing" long after the creator of the thought pattern has changed consciousness levels. In other words, one may die, but what has been created in the way of positive and negative thought forms continues to live, influencing each individual continent as well as the planet at large.[4]

The release of the built-up negativity of thousands of years of patriarchy will manifest most strongly at the stress points, the faultlines, where the tectonic plates meet. Young predicts bizarre and erratic weather, earthquakes, tremors and volcanic eruptions along these lines of stress. She also describes an east-west band from the 27th to 38th parallels north, as an area of highly unstable energy with potential for devastation. This band cuts horizontally across the United States; on the east coast it runs north-south from Washington, DC to Miami, on the west coast from San Francisco to Baha, California, in the central zone from Denver to New Orleans. The band is worldwide, running from the

27th to 38th parallels north across the United States, central Europe, north Africa and through Asia and Japan. These areas and the tectonic plate lines are the places where most earth change activity will occur.[5]

Young describes power as the central issue in the earth changes shift in human awareness, and speaks in "harm none" terms.

> No case can be made for the prima donna, singularly or collectively, who feels his whim is provocation enough to justify any infringement on another. The individual has the right...to pursue his own path, creating his own destiny, but this has never — and will never — mean that it is acceptable or allowable to move in a direction that causes another pain. The Earth must and will move past the impertinent and heavy-handed use of power into the refinement of power as a creative life force. This will happen, but not until man has unequivocally accepted the need for harmony and peace among his kind.[6]

Women have already learned this lesson, hopefully, which is a sending out of "harm none" energies.

She describes the direction of the years from 1988 to 2013, the time period most psychics designate for the earth changes to occur:

> The twenty-five years will be filled with increased fear and bloodshed as people struggle through war and natural disaster to reach a common vision. Wars will be wars of desperation, creative against static, vision against rhetoric. The slowly metamorphosing planet will suffer all the anguish inherent in being reborn, struggling to exchange the inappropriate and nonproductive for the intrinsically satisfying. As the general consciousness shift gains momentum, the turbulence of Planet Earth will gradually subside and the physical countenance of Earth will become one of calm. The realized focus of planet Earth will eventually be harmony among all living elements.[7]

That harmony is the New Age of the twenty-first century.

Meredith Lady Young's analysis touches upon some issues of world society right now. Her discussion of the ionosphere is particularly pertinent, considering the ecological crisis of the destruction of earth's ozone layer by basically selfish misuse of resources. The ozone layer is being damaged by CFCs, chlorofluorocarbons, released from refrigerants and styrofoam, as well as by carbon dioxide build-up from fossil fuel engines, particularly the automobile. The problem is further compounded by the cutting down of world forests, especially the Amazon rainforest of Brazil. Trees absorb carbon dioxide and release much-needed oxygen in their metabolic process, just the opposite of human metabolism, and provide an antidote to it and to engine pollution. With the ozone layer thinned and with a hole in it, much of its protection has been lost. The earth's currently increasing weather aberrations are believed to be caused by this ozone loss, which is also responsible for the increase in atmospheric radiation and cancer rates. Pollutants are no longer being reabsorbed enough through the atmosphere because of the loss of trees; they are building up, and making a build-up of contaminants in the human body as well. For Young, this is the crisis situation that will determine the earth's survival or destruction.

Her next issue is thought power, the human ability to create a positive or negative world by its collective use. Since all that exists is made of energy, and energy is made of thought, "what you send out comes back to you." In a world of "harming none," what is sent out is positive energy. But the world as we have it is immersed in the negativity of selfishness and power-over, and this is what has to be released, peacefully or destructively, before a different order can evolve. The negativity of a patriarchal order is part of the geologic energy of the continents and collective consciousness. That consciousness change from patriarchy to women's Goddess "harm none" values is essential, and will happen in whatever way it has to. With the old released, the new can manifest, a new way of thinking about the world. The areas of the tectonic faultlines, and between the 27th and 38th parallels, is exactly where beginning earth change phenomena is already occurring, and where most of the predictions of destruction are focused.

Young offers the power of positive thought as a healing for the earth, to release withheld negativity and replace it with creative change. She uses earth healing meditations for both individuals and groups, along with consciousness and awareness. She suggests directing healing energy to the faultlines and to the unstable energy of the 27th to 38th parallels. Group meditation increases the positive power geometrically, but individual meditation is important as well. Here is her In-Earth Healing Meditation for Groups:

- Sit in a circle with eyes closed.

- Concentrate on projecting one's own healing energy into the center of the circle, letting it merge with the healing energy present from other members of the group. (Spend at least five to ten minutes allowing maximum energy to be generated by each individual.)

- Arrange ahead of time to have one person begin chanting or repeating a healing mantra such as, "love heals and balances all."

- Stop the chanting and mentally visualize a golden or white light of combined energy from the group spiral upwards and outwards to the specific band or fault in question....See the spiraling energy then descend to the physical plane, balancing the in-Earth energy vibrations in the designated areas.

- Each person then quietly repeats to herself, "I place my hands on the energies in this place reconciling in the name of the Universe any imbalance or imperfection which may exist."[8]

Use this meditation or similar ones on a regular basis, informally or in the cast circle, with groups or alone. I make an earth healing meditation, usually the one that begins this book, a part of most of my rituals. Love neutralizes existing negativity. Be sure to visualize only what you wish to happen, as thought manifests on the earthplane, and is the creative power of women and the Goddess.

Marion Webb-Former, in her automatic writing of 1985, presents a picture again of thought power and alternate possible realities. Her guide Moon's description of things to come is harsh, and places the causes directly upon "man." Among the possible triggers for planetary disaster are listed contamination/pollution of the earth, an atmospheric dust blanket from volcanic eruptions, nuclear winter following a nuclear world war, and a pole shift.

> The possible disasters are endless, you may choose which one you choose to experience. Man is already making these choices during the dream state and he will project himself into whichever reality he wishes to experience.
>
> In the reality where you now exist, the Earth will shift its axis due to the combined belief of those within this reality that something must occur to stop man from his headlong dash into nuclear destruction. When souls combine their thoughts with but one purpose in mind, anything is possible. Thoughts are the tools with which a soul builds its reality and when one thought is joined by many similar thoughts, then the ability to create is greatly strengthened. In your reality, when mankind sleeps, its combined wish for an avoidance of what it considers inevitable is creating,...a tremendous vibration which is slowly shifting the Earth's axis. [9]

Women can create a new reality and put a stop to "man's" destructiveness and negative thought power. As long as people feel that war is inevitable, or that any disaster is inevitable, their thoughts give the idea energy and make it happen. Use thought force wisely, it creates and destroys universes. Knowing that, women project healing to the wrongs of the planet and to unstable land areas, and create instead a different reality, a positive one. Use visualization, meditation, healing energy, laying on of hands. Combined with direct action for political change and teaching to raise consciousness, this positivity can create a women's reality and alleviate much suffering and chaos. When the changes are complete and patriarchal negativity has been fully removed from the earth's aura, this positive thought and creativity will be the basis for a New Age.

Moon paints a picture of major devastation in the reality of 1985 collective consciousness, ironically triggered by the quest to prevent nuclear war:

> When your reality comes to the brink of nuclear war, there will be such a surge in the combined desire for an avoidance that it will cause sufficient shifting to drastically change the comparatively calm environment which Earth now enjoys. Land masses will disappear under the oceans and others, which have been lost under the waves for millennia, will rise again. Earth will spew forth her fiery core and the resulting fires will destroy many cities. There will be such death and destruction that warring nations will forget their petty quarrels and come to the aid of one another. Some nations will be totally destroyed and those remaining will be greatly reduced in number. [10]

If a part of changing this reality is to eliminate the risk of nuclear war, women are accomplishing much to change it. The global peace movement of the 1980s, created and run by women primarily, has taken effect. Most psychics today believe that World War III will not happen, though local wars will continue. Similarly, we need to now work to change the other realities, clean up the ecology, send healing to unstable geologic areas, and impress upon the world order the need to live by "harming none." There is much to do. Success partly depends upon releasing the idea that destruction is inevitable, another either/or duality. The earth will change, and this is desirable, but how it will change is the question open to many realities and possibilities. Change in the individual is also inevitable — and can lead to good.

Webb-Former describes what will happen in a pole shift:

> There will be tremendous flooding in some parts of the world. Ocean beds will be forced up in some areas and this will create tidal waves of such magnitude that other land masses will be completely consumed....Active and inactive volcanos will begin to erupt and some will be covered by water even as they do this. Great clouds of steam will be formed from the waters vaporizing as they pour into the center of the erupting volcanos. The moon's course will be altered and this...will affect the oceans...causing more flooding. The hours of darkness and light will change because of the Earth's new position in relation to the sun and moon. If the shift is anything more than extremely small, then Earth's path around the sun will change and this will create even more violent activity within her core. [11]

A large portion of Mu will rise from the Pacific, while Australia, New Zealand and other Pacific islands will sink. Japan will be mostly destroyed, with the remainder becoming part of the risen Mu land mass. Antarctica's ice will melt, causing floods. North America's east and west coasts will be under water, with low-lying interior areas flooded, until the United States map looks like two large islands. Increasing ice flow will make Canada, northern Asia and Scandinavia uninhabitable. Parts of Atlantis will rise in the Atlantic ocean, but not in large land mass areas. England and much of Europe will be gone, with Africa left surrounded by water.

Those surviving these changes will have other challenges to face:

> Great climatic changes will cause many to leave their countries in search of more favorable conditions. Disease will be rampant and there will be new dangers created by man's technology. A few of the nuclear reactors and power stations will not have been shut down in time before the tremendous upheavals begin. Most will disappear under the oceans, but due to meltdowns, they will erupt under water, thus polluting the surrounding waters and the atmosphere to a certain degree. Those which are swallowed up by earthquakes will explode deep in the ground, but there will still be some contamination from them. Many nuclear armaments will be buried under ice or water and the new race of man, which will evolve, will want to keep them that way. [12]

Webb-Former sees Africa as the place of the New Age emergence, the new beginning of a different kind of civilization. Racial intolerance will end, and the new people will become a mix of all the races. Some Asian survivors will settle on the risen land mass of Mu, finding Mukulian artifacts and duplicating their ancient, matriarchal "harm none" culture. The two societies of New Africa and modern Mu will eventually merge, providing the basis for a new world that has learned from the old one's mistakes.

There are predictions of a false messiah, bent on world power in the guise of spirituality, and twisting spiritual teachings for his own ends. He will eventually be assassinated by one of his disciples after the pole shift, opening the way for a true spiritual leader, or true spirituality to prevail.

Earth is a living reality, which has become highly imbalanced by human abuse of its ecosystem. All the possible realities, says Marion Webb-Former, lead to disaster because of this.

> When man ruthlessly destroys Earth's wildlife and pollutes her very being, he is not just upsetting the ecological balance of the planet. That is merely the outer reality of a much greater truth. He is removing vast reservoirs of consciousness which are a vital part of her. The remaining consciousness becomes disproportionate, having an overabundance of man's essence within it. Again, I say that there must be order within the physical plane, and therefore, the source will inevitably set right the imbalance within Earth's consciousness. That is why I have told you that for all possible realities there is some form of coming disaster, for in all possible realities man has created an imbalance in Earth's consciousness by some form of folly. [13]

Whether these things will come to pass remains to be seen, as conditions have changed steadily since 1985. The growing feminist influence of Goddess/women's "harm none" values is a major positive factor. Changes are being made in mainstream awareness and consciousness by this growth of women's and New Age consequence. Peace was a major women's issue in 1985, and still is, but there have been significant and rapid changes in world attitudes and structures since that time. Most people today do not believe that World War III is inevitable.

Environmentalism, which was not general awareness in 1985, is prominent today and both governments and individuals are effecting changes in how humanity treats the Goddess planet. The growing feminist, New Age and Goddess movements are teaching creative and positive use of thought power as a way of changing reality and doing healing. Much work on a spiritual and physical/political level is being done to create peace in the world, raise consciousness about power issues and "harming none," and to heal the planet and individuals.

Earth change predictions of the mid-1980s were particularly graphic and terrifying. Perhaps this was needed to prepare the way for immediate change, to force home the idea that the clear and present danger is now. Asked in 1990 if the predictions have changed, Marion's guide replied, "As always, my sweet, all is choice." More on her later predictions will be discussed in the next chapter.

Perhaps the strongest and most graphic predictions of this timeframe come from the Native American path of Mary Summer Rain in her books *Spirit Song* (1985) and *Phoenix Rising* (The Donning Company, 1987). Native American prophecy has long predicted a time of earth change, a time of the ending of the fourth world. There are traditions for this among many tribes, including the Iroquois, Sioux, Inuits, ancient Mayas and Aztecs, and the Hopis. The Hopis predict the ending to come via a nuclear World War III with China. The New Age or fifth world will emerge from a re-establishment of the Hopi ceremonials and a culture based on spiritual "harm none" values. Native people will teach this New Age to all who remain on earth after the changes and emergence.

Mary Summer Rain, a woman of Shoshoni heritage, was taught about the earth changes by No-Eyes, a Chippewa medicine woman and shaman. The beginning of the earth change teaching is described in *Spirit Song,* and greatly expanded in Summer Rain's later book *Phoenix Rising.* The material in both books is terrifying, but never without hope. It basically repeats Jochmans' seven steps in detailed perspective, using the metaphor of the birth and rising of the phoenix of the earth/New Age. The phoenix is a mystical bird that dies and is reborn in the fires of an ending world, a symbol of death, regeneration and rebirth. The entrance into the fifth world cannot come without the ending of the fourth.

In *Spirit Song,* Summer Rain is taken on a shamanic journey by No-Eyes, rising above the earth to see a redesigned planet.

We surveyed the entire continent of North America. The familiar geographic shape was changing its format at a rapid rate. We watched the movements as though we were viewing it through a child's ever-changing kaleidoscope. Yet this was no game. The earth was no toy. This was real. Everything had tilted down to the right. Alaska was now the tip of North America. Mexico wasn't south anymore, but rather west. New York was only partially visible.

On a closer inspection, we found that all of North America's east and west coasts were gone. Florida appeared to be ripped entirely off the continent. The major fault lines had cracked. The San Andreas ripped through the land like some giant tearing a thin piece of paper. The torn shred drifted out into the churning ocean....Hawaii was gone completely. Borneo, Sumatra, Philippines, Japan, Cuba, United Kingdom — all vanished with a blink of an eye.

We closed in on the United States. It wasn't as wide anymore due to the lessening of coastal areas....Michigan was covered with angry rushing waters. Upper Michigan had been torn away with the force of Lake Superior's emergency exit. All the Great Lake waters were forging downward, following the Mississippi River. Massive land areas on either side of the Mississippi River were flooded out of view....There was no land under an imaginary line from Houston to somewhere around Raleigh, North Carolina. Most of New York, Pennsylvania and Ohio were under water. *All* of Michigan and Indiana were. The United States was now divided completely. The eastern portion was an island.[14]

No-Eyes takes Summer Rain closer, to see the people, mob rule, a land in chaos and hysteria, urban destruction. Returning to the present, she gives a time of sixteen years for the pole shift (the year 2000), and sooner for a limited nuclear war to trigger it. The war is to come from Africa, and be stopped by extraterrestrial Watchers taking care that earth not be destroyed. The even limited nuclear exchange would set off gas pockets at the unstable faultlines and cause the pole shift. This was what happened in the 50,000 BCE destruction of Mu. According to No-Eyes, the pole shift and destruction, even the war, are inevitable. The spirits protecting earth are no longer sheltering its people from the consequences of their actions:

> It time for peoples to see stuff they make with thoughts, stuff peoples make with way peoples live even! Summer, *peoples* make all that stuff happen. It be peoples own effect![15]

Phoenix Rising, published in 1987, is a detailed expansion of the material in *Spirit Song.* Both should perhaps be classed with the earlier work of the last chapter, as No-Eyes gave the information to Summer Rain in 1982, but it was five years before it went into print. The metaphor of *Phoenix Rising* is of the earth in labor, giving birth to her cleansed and regenerated self, to a new earth in a New Age world.

Mary Summer Rain puts the progress of the earth changes, the movement from the old patriarchal order to a women's New Age, in ten steps that also describe the birth of the phoenix from labor contractions to flying free. A number of the steps have already begun and they are easy to relate to daily life. The ten steps are as follows:

1. Contractions/Economic Aspects
2. Emerging/Natural Disasters
3. Eyes Searching/Transportation Accidents
4. Listening/Freak Deaths and Accidents
5. Breath Heavy/Discord Between Nations
6. Wings Flexing/Spiritual Unrest and Awakening
7. The Wailing/Nuclear Incidents
8. Talons Tensing/Civil Unrest
9. Crouching/Massive Revolts and Government Turnaround
10. Flying Free/Rise of the Age of Peace[16]

Here is a summary of the steps and happenings predicted, with some current analysis. Some of the steps will happen simultaneously, and not necessarily in the order given.

1. The first birth contractions of the phoenix are to be evidenced by economic change and a collapse of world economies. As focused on the United States, this is predicted to mean an upswing in employment followed by

massive blue-collar strikes. The upswing is a surface appearance, and temporary, and the strikes will cause a dive in the economy. Striking factories and businesses will close and not reopen, leaving workers permanently unemployed. Management will take the operations to other countries, rather than give in to worker demands. This will result in increasing unemployment and Wall Street in confusion. Investors will sell out, causing the stock market to fall, and banks will be overdrawn. When individuals attempt to withdraw their savings in a panic, the banks will not be able to pay, despite the claims of government insurance. Recession will become depression with loan money scarce, causing small businesses to fail and new building to decline. Real estate will devalue.

The welfare system will be completely overburdened with the number of unemployed to support. To pay for it, national programs will be implemented rationing food and gas, and taxing foreign imports and domestic wages increasingly. Those working will be ever harder hit for taxes.

Some of these predictions are in evidence in 1990, with increasing strikes and increasing numbers of companies closing or relocating outside the United States. A company hit by strike or contract disputes forces concessions on its employees and unions, or closes altogether. In Pennsylvania, most of the steel mills that once supported the area's economy are shut down. The Volkswagen assembly plant is closed. There is a hard core of unemployed steelworkers in the area in the past fifteen years, and this is only one small segment of the country. The stock market has experienced two major panics since 1987 (October 19, 1987 and October 16, 1989). Junk bonds are a feature of Wall Street newscasts, along with corporate raiders who take over unwilling companies and run them into bankruptcy. Government intervention in the many savings and loans failures is costing billions of dollars, and probably cannot continue. Only the increasing inhumanity of the welfare system, offering less than survival and cutting off thousands of starving people, has prevented the collapse of that system so far. Despite the talk of a booming economy, there may be more than a million homeless people in America.

2. The emerging of the phoenix is accompanied by natural disasters of increasing frequency and magnitude. Stress on the Pacific plate faultlines continues to build and will release, causing earthquakes both in areas known to be unstable and in areas that have never felt earthquakes before. Volcanos that have been dormant will become active, and volcanos will emerge in new areas. Flooding will become an increasing threat in all coastal areas of the country, and the Mississippi River will increase in width dramatically to accommodate the Great Lakes emptying into it. There will be swamps where now is arid land. Hot winds will create dust bowls in other areas of the country, evaporating artesian water and creating sinkholes which will absorb residential areas. Evaporation by wind will erode fertile farmlands, and the high winds themselves will cause devastation and deaths. Tornados will increase, and hurricanes will happen with

increasing frequency and destruction. Even summer thunderstorms will become dangerous with intensified hail and lightning. Winters will bring record cold and blizzards, with week-long power outages and fatalities. There will be seen a greenish hue in the sky that is called The Phoenix Days.

Virtually all of these predictions are occurring in 1990. Earthquakes are increasing in frequency and destructiveness, with hardly a day gone by that does not report another one somewhere. Quakes are being felt in places that never knew them, as in Baltimore and North Dakota. Inactive volcanos are becoming active in Alaska and Japan. Flooding has been serious in the United States in 1989-1990, as well as in Bangladesh, Australia, England and Brazil, leaving many thousands homeless. In the spring of 1990, floods occurred in Florida, Georgia, Alabama, Texas, Arkansas, Oklahoma, Louisiana, Kentucky and Ohio, the reportedly worst in sixty years. Wind destruction is increasing in general and gradually becoming an issue nationwide, and combined with droughts worldwide is creating dust bowls, food shortages and famines. Soil erosion is a major world food problem. Tornados in 1990 in Kentucky, Kansas, Iowa, Oklahoma, Colorado and Indiana left whole towns destroyed and thousands homeless, and caused deaths. Hurricane Hugo (September, 1989) devastated Puerto Rico, St. Croix, Guadalupe, Monserrate Island and Charleston, South Carolina. Three-quarters of the buildings on St. Croix were destroyed, and 50,000 people in Charleston were homeless. The winter of 1989 broke 125 temperature records for cold in the United States, while the summers of 1987 and 1988 had record highs in the 90°-100° average range.

3. The eyes searching of the birthing phoenix bring transportation accidents, as the Watchers no longer protect earth from its unawareness. The accidents will be caused by simple carelessness, mechanical oversights, miscalculations, fatigue, human error, facilities in disrepair, etc. All forms of transportation, from spacecraft and airplanes to bicycles are included in this prediction of freak and preventable disasters, with carelessness causing much damage and loss of life.

In 1990, this prediction too is coming true. The Alaska Oil Spill of March, 1989 polluted 500,000 square miles of pristine coastline because of the intoxication and/or miscalculation of ship crew members. An increasing number of oil spills in oceans and waterways implicates carelessness too often, and destroys ever escalating amounts of wildlife and clean land and water. The January, 1987 explosion of the Space Shuttle was caused by a problem that scientists were previously aware of but did nothing to fix. In January, 1990, a jet from Colombia, crashed in New York City, leaving 72 dead — it was out of gas. Other plane crashes and near crashes have implicated mechanical failure, careless or otherwise. Train derailments create any number of toxic chemical spills, causing evacuations of whole towns. I have not even mentioned automobile and truck accidents. Interestingly, some of the major strikes are involving transportation — Eastern and Continental Airlines, and the Greyhound bus company, specifically.

4. The listening phoenix hears great suffering, from countless freak deaths and accidents. "They are hurting each other and they are injuring themselves." [17] The pressures of the earth changes will cause minds to crack and an increasing incidence of violence, murders and suicides. Freak accidents involving all sorts of machinery, from amusement park rides to chainsaws, will escalate. Old plagues and dis-eases thought long wiped out will return, along with new ones. Bubonic plague is mentioned, transmitted through rodent fleas. Four epidemics are predicted, two of them coming from biological warfare or research sources. One epidemic is credited to an accidental laboratory leak, and the other to a deliberate one.

Again, there is too much here that is familiar in 1990 as daily news truth. Bizarre accidents and senseless violence are on the upswing. In Pittsburgh alone, there was an 18% increase in violent crime for 1988. A soccer game in England resulted in 100 dead from violence. In Montreal in December, 1989, a man shot thirteen women dead and injured others in a college classroom because he said he didn't like feminists. He suicided, and the incident only made American newscasts for one day. A schoolyard shooting in Stockton, California, injured and killed a number of children in March, 1989. Airplane bombings, like the one over Lockerby, Scotland, in December, 1988, killing 258 people, are an ever-present threat. Senseless violence is a hallmark of our times, featuring terrorism, drugs and a horrifying array of weapons. For an example of bizarre accidents, four people including two children were killed near Pittsburgh in June, 1990. They were taking down a flagpole to repaint it, and the pole touched a power wire electrocuting them instantly.

The predictions of dis-eases are especially disturbing, considering the advent of AIDS. Most AIDS activists believe that the dis-ease was manmade and triggered deliberately or accidentally. The virus is universally fatal and its infection rate is escalating, destroying whole populations in some countries. Uganda and Zaire report as much as 85% of their populations infected, a whole generation dying out. In Roumania, infants are dying of AIDS. Their mothers are uninfected, and reuse of needles and contaminated blood supplies is the source. In the United States, what was once called the "gay plague" is now increasing rapidly among heterosexual drug-users and their mates and children, with two million infected people in the country. Infants are the fastest growing AIDS population in the United States, and most infected infants and women are Black or Hispanic. If there are three other dis-eases to come, we are in troubled times indeed. Cancer may also be considered here; one out of three Americans now dies of it. Of returning old dis-eases, tuberculosis is on the increase since 1988, and pertussis (whooping cough) has increased despite a vaccine. There was an outbreak of typhoid in 1989 in New York State, and cholera in 1991 in

South America.

5. The birthing phoenix breathes heavy and the nations argue. Blame, aggression and false accusations cause discord in the world. Economic collapse affects all countries as the world is interdependent, whether nations cooperate or not. There is depression in one nation after another, with embargos on each others' imports causing rising dissension. Nations intensify arms productions, to boost employment and for saber rattling. To justify the weapons, they increase propaganda and international disputes. Terrorism continues along with espionage, intrigue, and more hostilities. Small undeclared wars continue taking thousands of lives. General underhandedness is the hallmark of international politics, and the people become justifiably distrustful of governments and leaders.

Too much of this has been familiar through the 1980s, but here there is hope. The coming to power of a liberal government in Russia has ended the Cold War, at least between the Soviets and West. Communism has fallen in Eastern Europe, primarily because of food shortages and economic collapse (as predicted). A new form of government based on Western democracy is beginning in several formerly communist countries. There is depression, but also cooperation, and finally a decrease in arms productions and build-up (with much more needed). The situation in the Far and Middle East is not as hopeful, with repressions in China and increasing isolationism there, and continued violence, war, fanaticism and terrorism in the Middle East. South America is a hotbed of undeclared wars and foreign involvements, with too many deaths. The dispute over funding for the Nicaragua Contras and the world espionage of Irangate nearly toppled the regime of Ronald Reagan.

6. As the new phoenix/Age flexes her wings, there will be great days of questioning. This begins with the ending of separation between church and state, and a time of patriarchal religions forcing themselves on individuals and into government policies. There will be attempts to return religious repression to mainstream life, with legal persecution of nonconformists. This is a time of right-wing ignorance and paranoia. When governments have enough and there is turnabout, there is a reverse of affairs, with governmental control of religions and less religious freedom. Churches will no longer have nonprofit status and will be taxed heavily. They'll be held under strict government control.

While this is occurring, there will be an increase in UFO sightings and a new beginning of intergalactic contact. People will be infuriated at the longtime government cover-ups of alien existence, and there will be questioning of religious creation doctrines with a mass re-evaluation. There will be an almost universal reawakening of spirituality, a search for the real truths of human existence. The psychic senses will be accepted and validated, and psychic development taught in the gradeschools. There will be mainstream acceptance (or reacceptance) of reincarnation and karmic theory, and of the reality of life beyond the physical body.

Again, there is evidence in 1990 for these predictions manifesting. The right-wing in America is already too vocal, with representatives in government forcing religious "morality" on people's free will. Women's freedoms are increasingly in jeopardy in conjunction with this religious interference, along with other civil rights. There is hype about "satanism," used to sensationalize positive earth religions. There are attempts to prevent the National Endowment for the Arts from funding work that is outside fundamentalist values, and which the right-wing labels by their standards. Yet there is no corresponding crackdown on actual pornography, a women's issue and a male patriarchal pastime. Right-wingers in government predictably block laws that protect civil rights, freedom of expression, gay or disability rights, or women's rights. Two presidents have attacked abortion freedom and created an increasingly conservative Supreme Court. The line between church and state is becoming fine indeed, in this country and abroad. Right-wing muslim governments in the Middle East create a highly unstable economic balance for the oil-using world. One hope here is the debunking of the televangelists, whose exposed greed made many believers look twice at fundamentalist religions. Add that to the high public sentiment against Arab-world craziness, and perhaps the trend will stop before it gets more out of hand.

The end to cover-ups of extraterrestrial contact is long overdue, and most people even now believe in life on other planets despite ongoing government denial. It is a positive prediction of the earth changes that such cover-ups will end and earth will return to the community of intelligent worlds. The evidence is overwhelming that this planet was settled from other stars, and some of this truth may finally be forthcoming. If it causes people to re-evaluate patriarchal creation doctrine with its male birth, so much the better. The churches have kept the truth about human life from their followers for entirely too long. That psychic development and real spirituality will become a subject for common education is also long overdue. Reincarnation and karmic theory are in this category, also. With these concepts mainstream knowledge, patriarchal religions will have far less hold on the masses, resulting in a general increase in human spiritual evolution.

7. When the phoenix utters its first cry, the nuclear times will begin. Nuclear war is not the only topic here, but also the already clear and present danger of nuclear waste dump sites, unsafe transportation of radioactivity, and nuclear power plants. There will be earthquakes at two dump sites, and thousands endangered in other places by buried radioactive wastes leaching back up to the surface. Says No-eyes:

Earth Mother sick and tired of peoples burning waste in her breast. She gonna give it back to peoples. She gonna show peoples how bad she hurt. She give peoples back their own bad medicine even.[18]

There will be a nuclear power plant near-meltdown, and damage to the earth and people from chemical or radioactive weapons detonated by accident in an earthquake. Governments and authorities will continue to ignore public opinion until it is too late.

This can only ring as a true danger. In 1979 the Three Mile Island nuclear power station underwent a near meltdown, causing evacuation of the Harrisburg, Pennsylvania area. Incidence of birth defects in humans and animals, stillbirths, and high rates of cancer there continue. The 1986 Chernobyl accident in the Soviet Union made many more people aware of the dangers of nuclear power, but again public authorities ignore real dangers. In December, 1989, the Frenoval plant near Cincinnati leaked radiation into the public water supply. And in March, 1990, Three Mile Island again experienced leaks that amounted to a Stage I emergency — which was never made public. The increase in toxic waste transportation accidents could involve radioactive substances at any time. There is still no safe way of disposing of nuclear wastes — but the government wants to use them for food irradiation. Despite common sense and years of protest, the Diablo Canyon nuclear plant was approved to open on an earthquake fault in California, and the Seabrook, Massachusetts plant has been given an operating license. Public opinion and outcry have been ignored. A nuclear war is not the only nuclear threat to the planet.

8. The phoenix tenses its talons and the people riot against their oppression. The revolt is described as a highly destructive one but with little loss of life, and the end result is civil rights and equality for everyone. The Supreme Court is implicated in the passing of oppressive laws until the people can stand no more. This is the time of the nuclear exchange, possibly to be caused by a tragic mistake. One missile will strike New York City and the other a nuclear arsenal.

> Cities were in shambles from crazed rioting crowds. All minorities were taking revenge upon the unjust world. Buildings were crackling as hellish flames hungrily licked and gutted. Windows were blown out. Explosions ripped. Groups of angry weapon-wielding gangs roamed the rubble-littered streets in search for someone to take their madness out on. Sirens pierced the deafening din. Guns cracked. Laughter of the insane rang in my ears....I saw that I had been trampling through blood, blood bright and thick. I swung around just in time to see it all vanish within the blinding flash of a mushroom cloud. And I released all my remaining energy in a soundless scream.[19]

Anyone who has lived in an inner city ghetto in the heat of summer can believe that the possibility for such rioting is real. When there is increasing oppression and overwhelming poverty, something small can set off a chain reaction. Nuclear exchange by fatal error is not a new idea; there have been many near misses already. According to No-Eyes, the exchange will be limited by Watchers who stop most of the warheads from detonating.

9. As the phoenix crouches, the world rights herself in massive popular

revolt and government change. The people will refuse war and oppression, resist overtaxation and unjust laws. Arguments and shakeup in government will result in a major change of high officials and a reversal of laws and policies. Peace, democracy and unity return.

The United States government was designed as an antidote to oppression, overtaxation and foreign control. It seems conceivable that Americans will put a stop to oppression by their own government and take back this freedom. The nuclear exchange and rioting of Step 9 could be the last straw; change will come.

10. The phoenix flies free and the new order begins, validating human rights, free will, harmony and equality. Technology will be for the people and the earth, rather than oppressing both. There will be new nonpolluting energy sources and anti-gravity technology. World hunger will end, as will the abuse to human and nonhuman living things. There will be a universal free spirituality, peace on earth, and respect for all Be-ings and the planet. The Native American way of "walking softly on the Earth Mother" will be the way of life at last.

This is the New Age, however abstractly described here. It is a world of women's "harm none" values. The patriarchy ends, its technology-for-its-own-sake and oppressions gone, and a new era of respect for life begins. There will truly be peace on earth, and a government based on freedom, equality and spiritual values.

Says Mary Summer Rain:

> Perhaps that now we are aware of what is coming, man will uplift his mental attitudes, confront threatening forces, and make alterations and realizations as to his ultimate joint future. Perhaps the positive force of these corrective activities will indeed alter some of that which was to be. Perhaps man, through his new heightened awareness, may indeed circumvent some of the more devastating events. The advantage of viewing the future ramifications of present-day actions is, after all, a precious "gift of opportunity" to alter those negative avenues that would appear to bring forth harm and devastation.[20]

Perhaps indeed "man" will change his ways, and women's voices of common sense will prevail before more of the above predictions come true. Women's awareness makes a difference. We have consequence in the world, and by our changing ways of thought power can do much good. Perhaps a new order, a New Age can come that will right the wrongs peacefully, instilling women's matriarchal values without devastation.

Phoenix Rising does not mention the pole shift of *Spirit Song* directly. Nor does either book discuss ecological breakdown or pollution, outside of nuclear pollution. Mary Summer Rain and No-Eyes do not discuss the drug crisis as an aspect of international violence, or the oil situation of the Middle East. There are any number of possible realities and disasters, and any number of possibilities toward peaceful, positive change.

Notes

1. Earlyne Chaney, *Revelations of Things to Come,* p. 48.
2. Meredith Lady Young, *Agartha: A Journey to the Stars*, p. 172.
3. *Ibid.,* pp. 173-174.
4. *Ibid.,* p. 176.
5. *Ibid.,* pp. 178-180.
6. *Ibid.,* p. 151.
7. *Ibid.,* pp. 151-152.
8. *Ibid.,* pp. 180-181.
9. Marion Webb-Former, *On Earth Changes,* unpublished.
10. *Ibid.,* p. 2.
11. *Ibid.*
12. *Ibid.,* p. 3.
13. *Ibid.,* p. 6.
14. Mary Summer Rain, *Spirit Song: The Visionary Wisdom of No-Eyes* (Norfolk: The Donning Co., 1985), p. 144.
15. *Ibid.,* pp. 146-147.
16. Mary Summer Rain, *Phoenix Rising: No-Eyes' Vision of the Changes to Come* (Norfolk: The Donning Co., 1987), pp. 157-159, 146-155.
17. *Ibid.,* p. 149.
18. *Ibid.,* p. 103.
19. *Ibid.,* p. 111.
20. *Ibid.,* p. 155.

Current Earth Change Updates

1990

The earth waited, and still waits. We as a species have abused and misused and tortured our mother, our parents, our own true path and guidance. Still the earth waits, for the time is drawing near. It is time for harmony to return and give life and harmony back from where it was taken.

Tanith[1]

The earth change predictions paint a picture of a very changed planet, changed in disturbing ways. Earlier predictions make less attempt to describe what the changes are for, why they are happening and what women can do about them, than do later ones. The later the channeled material, the more emphasis is placed on what women can do and what is needed. Changes in individual consciousness, in group and in world consciousness are not only helpful, they are vital. These changes begin in the very basis of human thought, and in how women relate to the Goddess/planet, to themselves and to each other. In caring about others, women care about the earth, since all are Goddess. In caring about Goddess, women care about the physical planet that is her body. In caring about themselves, having a positive self-image and healed emotions, women create a basis for both. "You are Goddess" is a central Women's Spirituality law.

Other Women's Spirituality ideas are vital in creating positive change on the earth. "Harming none" is basic, creating a world where everyone is free and equal and respected, a world where everyone's real needs are met. "What you send out comes back to you" is another idea, affirming that positive thought creates positive change. By thinking that only disaster can change the world, only disaster can. By thinking that change comes in positive, peaceful, transforming ways, it does. "Be careful what you ask for" is another Goddess idea in regard to earth changes; women asking for and visualizing earth healing and peaceful change help to create it. And one further idea, "You have consequence in the world," establishes the attitude of creating change and of taking an active part in revitalizing the planet. With these Goddess laws in mind, only good for all of life can happen.

In reading about earth change information, be aware of thought power.

What you visualize is what you create or ask for, and it will come to pass. Instead of seeing the predictions of destruction happening, see what the possibilities are of changing the destruction into positive energy. Instead of visualizing a nuclear accident, see both the situation and how to prevent it. By seeing both sides, what there is and what could change it, women realize the gravity of the situation without lending thought power energy to manifesting it. When the women of England realized the danger of nuclear war and of nuclear weapons bases on their soil, they created the Greenham Commons Peace Camp. Without denying the reality, they put energy into change and were successful. Their efforts brought to world attention — and stimulated world protest against — the patriarchal insanity of nuclear war.

Women reading about earth change predictions do the same. They recognize the reality and gravity of the present situation, and put energy into changing it. Earth changes, if they happen, are the Mother's response to five thousand years of patriarchal abuse. Women who have the awareness to understand what is happening are the planet's hope for solutions and survival. Realizing that severe weather is a potential source of disaster and loss of property and life, visualize the winds gentling, the storms easing, the tornados moved away from populated areas. See the earth and her people and animals protected. When a trouble spot is known, send love to heal it and to lessen the pain. Where people are injured or homeless, again send healing energy, along with physical aid. Put your hands on the earth often and offer her energy for peaceful evolution. Do the activism that causes political change. Be aware of the needs of the environment and help where you can, physically as well as in energy channeling.

Most of all, women's changing awareness is a force of positivity in the healing of the earth. By saying no to environmental abuse, to oppression of peoples, to patriarchal greed and negativity, women create change and raise others' consciousness so that more and more people create it with them. By protesting as publicly as possible the things that need changing — from apartheid to world hunger, from discrimination against lesbians and gays to the oppression of women — public awareness is raised and positive change begins. By removing oppression and poverty from the changing world, perhaps the rioting won't have to happen. By honoring workers' needs, perhaps economies will grow strong instead of fail. By changing one link or many in the earth change chain of events, many negative happenings are prevented from needing to run their course. If the earth changes are a response to world and human needs, by meeting these needs the physical manifestings are lessened. In raising human awareness, women have the potential to create changes with world consequence.

Reality is what women create with their power of thought. Disaster is not inevitable, but the changes to bring about the equality and peace of the New Age are. The changes can happen chaotically or not, there is choice. How women create reality, what they visualize for the present, the changes and the future, is

102

important. By full awareness of the danger areas and of what is predicted to come, women's ability to apply thought can change or lessen the reality. By knowing what needs changing, women's creative power can know what's needed and what to do. By having as much awareness of what needs healing as possible, and awareness of the current trends and predictions, women use the information to place their healing and consequence in the best possible ways. By visualizing clearly and specifically what a women's New Age can be and what a new world will look like, women can create that world in ways that are positive for all. Magick is the art of creating reality at will; women create wisely.

The patriarchy has given us a status quo that is destroying the earth and many of her people. A world run by white, middle-class Western males is not a positive world, even for the patriarchs who run it. Women's potential reality is Goddess and earth-based, life-affirming, and validating of all peoples. There is no discrimination in it against any race, ethnic group, nation, religion or sexual preference. There is no discrimination. People have the right to be who they are, who they choose to be, without interference. In the rule of "harm none," all are free to follow their own good and own needs, as long as they don't interfere with the free will of others. When this rule is followed, everyone's actions are for their own good, and their own good is also affirming of the good of all. A matriarchal/Goddess culture is a culture that values all life, everyone's life, be they two- or four-legged, and the life of the planet herself.

This is what we are working toward, and this is the reason that information on earth change prophecy is important for Women's Spirituality now. By being aware of the problems, by knowing where current patriarchal abuses are leading the future of the planet, and by understanding what a New Age can be, women have the tools of change in their hands. Using thought power and imagination, women's consequence and understanding, women's willingness to do healing and create change, the New Age can come about in the most positive ways possible. Current predictions all say that destructions will happen, and they are already manifesting. How destructive, how much loss of life, how disrupting they will be is open to working for change. By using women's awareness and both psychic and physical action, the New Age can happen more easily, and be the best we can create. The end-result predictions for the earth change era are a thousand years of peace, and an Age of Women.

On Earth Day, April 22, 1990, Marion Webb-Former asked Moon "Will the 'New Beginning' still happen as you predicted in 1985?" The response speaks of the changing awareness on Goddess Earth:

> Within the illusionary time span since I last wrote of these happenings, many enlightened souls have been attracted to your quadrant of the physical universe. Some are incarnated upon Earth, many are dwelling within your galaxy....Certain souls on Earth are in contact with both individual and groups of watchers and creators who reside within your galaxy. Some of these souls have shared their

channelings...through what you consider the "media" and thus, their words have reached many. Others, who remain unknown, go about their physical existence sharing their enlightenment by thought, word and deed. When examined closely, these individual incidents of enlightenment seem intriguing and thought provoking, but of no great consequence. Yet, when viewed from beyond the realms of the physical, the pattern becomes clear and the increase to the vibrational flow is evident.[2]

Women's Spirituality is female, and two-thirds to three-quarters of New Age and wiccan participants are also women. It is therefore women's "enlightenment" that has created the consequence Moon describes. By our changes in awareness for the earth, ourselves and others, women are creating positive change. According to Marion and Moon, the threat of nuclear exchange is lessening and that of a pole shift has decreased.

There has begun the possibility of a "shift" within my predictions of an axis shift. The potential for this occurrence continues to be, but due to the gradual increase in soul vibration across your planet, the possibility of nuclear confrontation begins to fade into the realms of other probable realities....

Do not, however, misinterpret my words, the "New Beginning" is coming, for it is timely, and great havoc and devastation will disrupt your planet as she progresses into a new millennium. A slight axis shift may yet occur, but may be more minimal than previously described. Within the remaining decade of what you perceive as the twentieth century, the ultimate decision will become a focus for the group mind. The progression is at hand for your planet and throughout the enlightened physical universe there is anticipation and rejoicing.[3]

Writing of Earth Day, which brought response from hundreds of thousands of people nationwide, Moon comments:

As evidence of the increase in the vibrational flow of your planet, consider this day and its meaning. At last, a true awareness of man's destruction of his home planet is just beginning. Celebrate this day and project healing thoughts to all its inhabitants and all its elements. This day holds the potential of increasing each soul's vibrations, as did the Harmonic Convergence. Treasure it and know that it is yet another opportunity for enlightenment, a new threshold opened up by the source for all souls who desire oneness to pass through.[4]

Tanith, in her superb readings based on the tarot, writes about the earth changes in May of 1990. She feels that great change is possible and will come, that there will be disruption and pain — and an end to patriarchy:

We have had time to learn lessons, but have refused to see what we must do. Instead we wallow in our conceit and unknowing, blinded [sic] to the lessons we must learn. The air is dirty with our misdeeds. All await the changes.

And the changes come. They are led by those who love, and have already started the journey. Many are the women who now deal from the heart, not the mind. For so many years we were told not to be emotional or listen to that silly women's intuition. Now those that know, those that remember, are listening to the water within them, to the earth within them, and responding with the fire within them. The changes will not come through more rules and laws, for these are things of the air and thought and mind. They are not coming through committee meetings or press conferences. The changes are coming through the hearts of those brave enough to love, and to send that love outward. These are women who act because they need to, because they are called to the harmony that once was. [5]

Women are the leaders of the change from patriarchal mind to matriarchal heart. Each in her small way has great consequence:

Will there be rulers in this time? Yes, rulers of love, of the heart, of their own direction. They will be leaders and teachers, not kings. They will be priestesses of small groups and healers whose touch spreads out and permeates even the earth itself. Healers are the way of the future, for as they heal each other, they heal the harmony and the earth.

What will happen to men in this time? Some will understand and join the love that is surrounding them. They will protect and defend with love. Some will not understand and they will band together and create fierce rituals of war. They will destroy what they see, but will eventually destroy themselves. With them will be some women who only understand power over, not the power within, or the power together, or the power of the Mother. This again will be the war of water, earth and fire against air. In some cases there will be a truce, a peace. In some there will be war that destroys all that is near. [6]

Tanith sees no world war in this time of change, but wars will continue. She sees physical devastations, and changes in patriarchal governments. She sees a new world of healing and laughter to come:

I see no great world war, however, local wars that kill those near to them. Wars that can be left if you so choose. This in part is the Mother letting those that cannot deal with harmony leave this world. Perhaps there is another one for them.

We will fear the fire in these times. There are drastic earth changes coming in the form of drought and flood; earthquake and earth-sinking; sickness and ineffectual medicine. These are messages to the children of the earth. Too many of you do not heed the warnings and signs. Those of you who do not listen must be shown. Listen well, for only those who listen will survive. The earth is your Mother, she cannot exist if you continue to abuse her. She will set the change in motion, but you must listen.

The changes set in motion will disrupt society as we know it. The governments will not be able to fight, or argue, for the people in their supposed care will be calling to them. Some men at the top will continue to act as if they rule, but no one will listen. Smaller groups will spring up to fill in the gap of the larger ones. And yet, eventually these smaller ones will join together in one great large network, a women's network, run by women who understand. Healing and laughter will eventually be the concerns. We will forget the need for more land, more money, more toys.

There is a battle coming between the large and the small, but the earth will help us, for she wishes to heal. Many will not survive the changes, but those that do will begin a new way, an old way, the way of harmony. [7]

At the end of the trauma is hope, consciousness and regret, and rebuilding in a New Age. Harmony on the planet is harmony among people, animals and the Goddess earth.

We will all cry over what we have done. We will see the terror we have caused as a group, as a race, as a species. We are all to blame, yet we are all responsible for the change. The change will be hard for some, but in the end good for the family of creatures and the Mother herself. [8]

On May 19, 1990 Tanith wrote the following:

The increase in volcanic activity, earthquakes, floods, droughts, high winds, heat and cold is sent by the Earth Mother to clean and replenish the planet. The increase in plane crashes and human self-destruction in the form of gangs, shootings and drugs is due to the need for a cleaning out of the species. Much as nature provides for the death of the weak of the species, those that will not help the species survive, nature is providing the same for humans. This is survival of the fittest in terms of healing, leadership, loving, psychic and spiritual abilities. These are those that will create the future. [9]

She writes again about the role of government, the abuses of governments that will end, the change from patriarchy to a new matriarchal order:

Governments will be forced to care for the members of their societies instead of fighting each other. There will be support for bigger weapons, for international espionage, for politics that interfere with other countries. Each country will be involved through necessity in the caring of the people that live there. Political battles will be useless and laughable. Men will fight for political advantage and rank, but women will gain the positions. The issues that men deem important will no longer interest the people of the land.

Many many women are going through great trauma at this time, probably beginning in 1987 with the August 17th Harmonic Convergence, and ongoing since. These are women's personal earth changes, reflecting the changes happening in and on the planet. Tanith discusses this, describing 1990 as a crucial year:

This is the great year of challenge and change, the year that separates those that will lead from those that will follow and those that will not be part of the future. All that is being experienced is to burn off the outer cloaks and covers and masks worn to hide the inner self. The masks must be lifted to free those that are needed. By the last turning of the wheel of seasons all that has been learned will be integrated into the essence of each of us. Many past lives are now coming together inside those of the future. Memories return from the past to lead us into the future. All that was forgotten will return. In order for the rememberers to receive and accept this information they must be emptied of their blocks and blindfolds. They must be willing to finally see, see with their entire essence. That is the challenge of this year. Some will face it and grow, some will face it and die, some will run away and stay as they always have been.[11]

The changes are happening in the now and are profound for both women and the changing earth. The women's New Age that Tanith and others predict will be described in the next chapter. For now, there is encouragement and hope for all who live in these times:

Fear not for the changes are coming and will benefit all the strands of the web, all that dance and move together in this interconnected world. And through it all the ancient ones will be here to love and guide us.[12]

Mari Aleva, in her earth change channeling session on February 5, 1990, describes many changes for earth's future, some of which are starting right now. Delta Four, one of the entities who speak through her, explains:

As we all know, the change is inevitable. Change takes many forms....The part that we feel you are interested in at this time is the group thought of earth here now. There are many dynamic group thought processes. Some are what you would base negative, some positive, and a lot in between. At one level we have your warriors, your gun crazies, whatever you want to label them. And at the other end you have your peace makers, and you have everything in between, including the so-called churches that bring the light of the Christ within and they harbor guns. Thought forms are creating new energies (and) as the light communities of the new age expand, do look for the other side....Does this answer your question?

Diane: It answers a lot of questions, yes. That even though change may bring destruction, the destruction is of the old and something new will come that could be a lot better. Am I interpreting, understanding that correctly?

Delta: Destruction occurs in many ways. Some destruction, what is termed destruction, is actually a rebirthing. As you destroy the old skin, the snake continues to live.

Diane: What about all the changes inside individual people that are happening right now, the type of destruction defined as death and rebirth that I am referring to? I see that as a major part of what earth changes means is the changes in consciousness in individuals.

Delta: As the consciousness changes, your earth evolves in that direction.

Diane: Is it possible that people's changing, the consciousness changing of people, can prevent the physical changes that are prophecied?

Delta: Changes always occur....Earth is filled with poisons made from humankind.[13]

Later in the session Delta says: "The survival of the earth energies are dependent upon the energy thought forms that humankind places upon the earth."[14] A poisoned planet will result in poisoned people, with an increase in illnesses and dis-ease. She urges the planting of trees to clean the air, but states that the poisons in the environment are also thought-form energies, the poisons people have put into the world by negativity. As insects become immune to bug sprays, until stronger and stronger chemicals are required to kill them, so is negativity in the environment:

Delta: People are changing in the same ways, and new diseases are coming.

Diane: I understand why they're coming, but what can we do to change this, other than to plant trees and to raise peoples' awareness about caretaking of the earth and about positive thought?

Delta: Thought form is the utmost importance. Do be prepared to realize the opposition.[15]

When Helm returned to speak through Mari, I asked again about individual changes, mentioning the Harmonic Convergence and the drastic energy shift in individuals it seemed to precipitate. If the changes in women are mirrored in the changes to the planet, women's internal changes become highly important in what will manifest as earth changes. She also talks about a pole shift:

Harmonic Convergence is a very interesting thing. It involves the planetaries, the planets, the different poles and how things were in a precise time, somewhat like when Jesay was born, Jesus you call. There were energy changes that occurred during this time. Many doors have been opened. You're speaking of much negative. Ah, but know as these things, the other side as Delta mentioned, you have ones that are growing.

Diane: So it's the release of negativity?

Helm: Many people you'll notice now opening. They come to you and they say, well a couple of years ago they do not know Harmonic Convergence, and they say couple of years ago I start getting into this thing....Most of them will say since 1987 or '88, '89 they have been getting interested in these things. Because of the energy change....

And another thing, you can put this one in your book, this is another one going to confirm if it is not already there. We do not know. The pole in the earth have changed a fraction of inches, very few. As that occurs, the magnetic things in there change and that does cause some disruptions around your earth....

Diane: But on individual terms, people are changing and that's because of the change of energy? They are growing?

Helm: Yes, what you would call universal world change.[16]

This individual and universal change includes changes in the world situation. When asked if people's changes could prevent negative predictions from manifesting, the answer was "absolutely." Here is more on the world situation and the predictions of a world war.

Diane: There are prophecies of a war with China. Is that a necessary thing, or can that be prevented?

Helm: We are not knowing. We were going to share about things happening that are opening up. Also watch that the thing open up here, and you don't know what the other hand is doing, 'cause you are looking at this wall come down here (Berlin Wall). Always know that there is another land somewhere. You mention China. How about India? What is happening with India?

Diane: I don't honestly know. They're developing nuclear power, which is kind of a concern.

Helm: You bet your bottom dollar they are.

Diane: Okay, and that's a threat.

Helm: The ones you don't always hear from, the quiet ones, might just surprise people. We are not wanting to think war. The more people who think war bring it into possibility. There are wars going on all over these countries. Some take in entire countries, some entire cultures, some are in religions, some in peoples' own where they work and some in their homes.

Diane: Will that change for the positive?

Helm: It is individual. It goes from large (world) to even inside people's homes and from there into their minds. When they change inside their minds and they open and feel better in their homes, then the communities get better and the countries and then the world. Starts from here.[17]

Change in the world begins with change in people, particularly with change in women. Mari and Helm spoke earlier about women being dominant in earlier earth cultures. What about women in the earth changes now?

Diane: I also understand that what part of the earth changes means is it's been named by some as an Age of Women. So what's coming for women as human consciousness changes, as the earth changes and we solve or not solve some of the problems of the planet?...

Helm: They are gaining in strength, and the men are gaining in fear. And out of fear comes some pretty erratic behavior. Women are gaining in their equalities, they have been.

Diane: Will they be able to hold on to that and increase that?

Helm: There are many levels. The men are feeling fear. When ones feel fear they do not always think wisely.

Diane: So further oppression can come from that. I think we see that in the politics.

Helm: That is correct.

Diane: Okay. What can women do to hold onto their gains?

Helm: Support each other. Support each other and open the mind and think positively and act in light, what they call good stuff. It is much, an ounce of love will take care of a pound of hate. But you need to project that ounce of love. It is much stronger than hate.

Diane: I also see some men changing, and some men becoming gentle and opening their own feminine sides.

Helm: That is correct. Women are becoming stronger. Men are becoming softer. And it is to more of an equality, like we say,...[18]

Mari and Helm see a balance of energies in the New Age, but "women are becoming stronger." This mirrors some of what Tanith said as well. From the work of both these contemporary women there is much to hope for. Marion Webb-Former's channeling also shows positive change. Through women's increase in awareness and its effect on the planet as a whole, the earth changes may differ from what was earlier predicted.

Laurel Steinhice is a conscious voice channeler from Nashville, Tennessee. One of the entities she channels is the Earth Mother, who names herself Mary, Isis or the Goddess. She radiates great love and nearness, speaking through many channels to assist the transition of the age. Being witness to this channeling was an experience of overwhelming beauty and great healing. The presence was clearly and definitely real. Laurel Steinhice is also the only public channel at present of Edgar Cayce, the great early healer, psychic and earth change prophet (1877-1945). It was Edgar Cayce's work that popularized the psychic world for the mainstream. Again, I can only affirm that the presence was a genuine one. Most of this book was completed by the time of her June, 1990, channeling session, or it would appear throughout. Read the whole session in the Appendix, it is highly important. Some highlights appear below and in the next chapter.

Here is the Mother, speaking through channel Laurel Steinhice, in a message that is probably the central idea of this book:

When you heal yourself and assist others with their self-healing, you heal the earth. For the earth is one with everyone and every creature. All are part of the earth. Those that creep and crawl and go upon many legs, those that fly in the air, those that go upon four legs or two, are the Earth Mother's children. All are part of the earth. The healing of any one of these is healing that contributes to the whole. [19]

I asked, as a healer, do women's efforts at healing really make a difference in the earth's changes?:

But you do [make a difference]. That is the message we bring. *It does matter.* The simple loving of the earth is healing. When you love the earth you bring energy for her healing. There is interconnectedness between earth energy and personal energy, and use of that energy is what the transition is all about.

Diane: May I ask in what name you are speaking to me?

Mary: I am Mary. I am also the Goddess, known by many names. You are familiar with my energies....[20]

Edgar Cayce, as channeled through Laurel Steinhice, is a leading authority on earth changes. Updates are coming rapidly, as so much has changed and been changing in the last few years. He elaborates on the idea that women have consequence as to what will come.

This planet is changing. One does not need to be psychic to see that. Changes around you everywhere, everyday. And it is changing so fast! The whole program is changing so fast. We bring updates from time to time.

By loving the earth, you help to heal it. Personal caring makes more difference than people realize. Let us give you some examples of this. In 1988 there was a great hurricane in the Gulf of Mexico. And everyone said, "This is the worst storm in a hundred years." Perhaps you meditated or prayed, or simply sat before your television set and said, "Oh, we hope this doesn't go ashore where it will do the most damage."

This caring, the positive energy of persons of good will everywhere, whether incarnate or discarnate, combined to push this hurricane away from where it would have caused the most harm. This is your success, and that of many....

In the same way, the negative energy which would come forth from the earth in violence, in a man-made or so-called natural disaster, is assimilated, is pulled out and harmlessly released into space by loving the earth....Yet, as you yourself have noticed, there is still so much work to do.

Lightwork reduces the likelihood of a negative force release in some other form. Think how the world peace movement has manifested world peace. And now we decided not to have that war at all. [21]

Likewise, the threat of high-destruction earthquakes is lessening. Balancing is being done of energies to release pressure with the least amount of damage. Earthquakes will continue, but not in the magnitude of the original predictions. However destructive quakes might be, they will be less than what would have happened if human awareness had not changed. The purpose of the changes is not to preserve life, but to preserve the actual planet. Says Edgar again:

Ultimately the purpose of this transition, the whole transition process, is not to preserve physical bodies. There are too many bodies on the earth. It is to make an orderly transition instead of a disorderly one. Human lives, human bodies, have been recycled for some time now. The preservation of the body is not the object. Is it the saving, the preservation of the soul? No, for the soul cannot be destroyed. The spirit lives forever, and if we lost this planet, what would we do? We would go somewhere else and start over. And it wouldn't even be the first time such a thing has happened in the vastness of the universe.

Is the question whether this will be a planet of darkness or of light? No, that question was resolved in the eras known as World Wars I and II and in between. It will either be a planet of light or it will be gone. The issue is the physical preservation of the earth planet itself. [22]

Earth in moving from the fourth to fifth world is also moving from a third dimensional reality to the fourth dimension, which will be the New Age. That change has already begun, with the changing of (primarily) women's awareness in the last few years. Laurel Steinhice and Edgar Cayce call this awareness "Christ Consciousness" but define it in new ways. They discuss the Harmonic Convergence in this light:

> That event some people have called the Second Coming, the Christ Consciousness manifesting, was originally projected for 1998, to be followed by the shift of the axis in 2001. The Second Coming manifested ten years ahead of schedule, in 1988. And let us say here that we're not talking about Christ as a male, we're talking about an energy called the Christ Consciousness, which is as much Earth Mother as it is any other energy.
>
> Diane: What does that mean when you say that that came to this planet?
>
> Edgar: It started in 1988, ten years ahead of schedule, the first stage of the project being the rising light consciousness in the hearts and minds of people.
>
> Diane: The Harmonic Convergence?
>
> Edgar: Yes! So, it is ten years ahead of schedule. That does not mean that the final shift comes in 1991. It means we buy time to do more work to buy more time. And the shift, instead of being in 2001 will perhaps be as late as 2020. Or perhaps as early as 2012. For we have been so successful in raising consciousness, all of us together,...that the healing is ahead of schedule, well underway. [23]

Thought power, consensus reality, is described again as determining what will happen before the pole shift, and how dramatic it will be — what will mass consciousness create? Before the shift, there is predicted a movement away from earthquakes to manmade disasters: "oil spills, chemical contamination, terrorism, random violence. All of these are already on the upswing." [24] The purpose of the shift is to purify the contaminants that are polluting the earth. They will be folded under the earth's crust and broken down there.

What does a pole shift mean in geologic terms? Will the earth actually turn herself over?

> There will be, at the final shift, in a single twenty-four hour period, a tipping and shifting of the axis. It is not actually the core of the earth that will move in relationship to the universe. The core will be where it is now. But the crust, there will be big bubbles in the core that comes and breaks and jolts loose the crust, and the crust will float around the core. And then reattach....And afterwards the magnetic flow that is now in the Northern Hemisphere will be in the Southern, and vice versa....
>
> This will be the fourth time this has happened. Ask the geologists. They'll tell you this has happened before. [25]

Geologists say that this has happened many times. During the transition, before and after it, the world will be much rearranged. These are the most recent predictions in a changing scenario:

> There will be a new bay in the Gulf of Mexico, where portions of South Texas, Louisiana, Alabama, Mississippi and Arkansas are now. Memphis is still expected to be a seaport. This is update, but it correlates with the projections that were given some time ago.
>
> Diane: So those cities will be destroyed?
>
> Edgar: Many, yes. But there will be warning. People can readjust. There will be a rising in the Gulf of Mexico, and right now California looks better than it has in a long time. We think California might be preserved. Not surely, but a good chance of preserving that part of the plate. Alaska is looking worse. The oil spill caused priority to fall for preserving resources in Alaska, resources being trees and natural resources. Therefore, some of the negative energy that would have focused on California is shifting to Alaska instead. And we expect seismic activity there relatively soon.
>
> Japan is still in danger, but is looking better than it has in a long time. You know the crust, the plates are such that one folds under another? Japan was expected to be folded not only under the water but under the Asian landmass. The only way to prevent this from happening, we have been in the process of working on for some time. It involves controlled seismic activity. We take Japan and we separate it from attachment to the crust, float it on the liquid just under the crust farther away and reattach it in a safer place.
>
> Every time this controlled seismic activity is triggered, there is the very distinct danger Japan will all go down. It is calculated risk. If we don't do it, it goes down for sure. And if we *do* do it, *maybe* it is preserved.
>
> Diane: What is the potential for preventing destruction — loss of life, major damage — by the light workers that are working toward changing the things that are wrong?
>
> Edgar: Excellent potential.[26]

The planet is changing dimensional frequency from the third to fourth dimension, and is already more than halfway into the New Age. This fourth dimension may also be described as the "psychic reality" level, or what in women's rituals is called the "place between the worlds." Those who are light workers, those with awareness, are already manifesting that reality in their daily lives. It is this increasing energy frequency that is causing the traumatic individual emotional changes and healings, but is also causing the violence and negativity. Negativity released, as blocks let go of from personal lives or even as random violence, is released from the planet as a whole:

> This is the stress manifestation of rising vibrational frequency which, no matter what path you are on, is an acceleration. The violence is increasing but these energies, when they go out, they don't come back. It's not the same way. They are being withdrawn from the system.

Diane: But they're still manifesting.

Edgar: Yes, because we are still in that stage. But they are being withdrawn. They are being shipped out....Everyone goes to the personal reality they have created for themselves, and not all of these are pleasant. But they are being shipped out. [27]

Personal reality also determines how the earth changes will manifest, and how much of the earth's negativity will be released by the time of the pole shift. These in turn determine what the shift will be like:

The shifting of the planet: it is either an orderly shifting or a disorderly shifting. Where it is disorderly the term destruction is quite apt. Where it is orderly, then there is less that you would perceive as destruction. It is simply renewal, realignment, readjustment. If everyone will listen and plan and participate in the orderly transition, then there will be no disorder, there will be no destruction. But not everybody listens at the same rate, the same pace. [28]

And when people hear, will they heed? Edgar through Laurel Steinhice talks about the last days of Atlantis and Mu, and the people who knew but didn't heed:

Hi-tech Atlantis — what happened? We knew the shift was coming and we told people, "Go to high ground." We knew where it would be, what was happening, and we said, "move now." And everyone said, "Oh, we'll wait to the last minute and then we get in our little air cars and air boats and we go for higher ground." But the transition, the shift, is geomagnetic in character — realignment of the polarity of each molecule. And approaching the shift, the geomagnetic grid was disturbed. And the air cars and air boats and all the full technology of Atlantis was geomagnetic based, controlled by crystals. So it didn't work. When the power fails, who do you blame? The electric company. We didn't do it; we just got blamed!

Now Lemuria, which is low-tech. Instead of saying, "we jump in our cars, we jump in our air boats and go for higher ground," they said, "we will teleport. No problem. We know how to do this thing. We will teleport." What kind of energy is teleportation based on?

Diane: Also geomagnetic?

Edgar: Yes! So that also didn't work. And it was too late to walk.

Diane: But they knew, they knew when it was coming?

Edgar: They knew it was coming. And some of them had sense enough to start walking early on. Or to teleport themselves before the geomagnetic grid was disturbed. But the essential reason why Lemuria went down was the same reason why Atlantis went down.

Diane: The culture?

Edgar: It was the shift, the polar shift, the realignment of each molecule, the electromagnetic force. It's what is happening to the earth again. And this technology will fail. But here again we say, it is not a question of whether the transition will come, it is a question of whether it will be orderly or disorderly. And new technologies are already being developed to fill the gap. [29]

Whether the change is orderly or not depends on human awareness, on a critical mass of conscious, aware people who will heed the warnings to move when necessary. After the shift, which will be the ending of a long process of preliminary changes, comes the rebuilding and the New Age. Laurel Steinhice and Edgar Cayce call it the fourth dimension, and there is more about it in the next chapter.

But what will happen to those who die? The soul is eternal and will move to other planes, other planets, other realms. Laurel and Edgar call those who leave out-transfers, and state that after the pole shift there will be only 10% of current population on the planet. Where will these souls go and why will they not be able to make the transition to a New Age?

> Those who are moving to new neighborhoods of the universe move for many different reasons. Some because they cannot get their karmic act together in time to graduate to a new dimension. They move to another dimensional planet where they will have the opportunity to learn and grow, each at his or her own pace. That is the choice, and it can even be a high choice to do this.
>
> Some will move to other fourth dimensional planets, positive fourth dimensional planes much like the new earth will be. They're moving to the same kind of neighborhood, but a different street address in the universe. And of course they don't go at random like seeds scattered to the wind. They go in groups of personal affinity — soulmate groups, oversoul selves, separate selves — however you want to put it. What it amounts to is if we are all going somewhere, we don't go stragglers one-by-one, we hire a bus.
>
> And some will go *home,* after a long time of having helped those who most need help. They are the workers, the troubleshooters of the universe, who will go home, have a nice visit and rest, and most likely thereafter say — "Well this has been wonderful. Now what's my next assignment?" For surely you don't think, in all the vastness of the universe, that this is the only place there is spiritual sentient physical life or that this is the only place that ever got itself in deep trouble. [30]

Only in patriarchal religions is death a thing to fear. Edgar talks of healing and death in the earth change era:

> Let us not overlook the fact that those who choose to discard the body may be making a high choice. Don't make them feel like second-class citizens. Healing is not always the preservation of the body...
>
> There is healing in which the body is still discarded. If someone has chosen to make the death crossing transition, for whatever reason, and that is true to their choice, your lightwork won't stop them. It will make them more comfortable in the process, and this is a wonderful thing.
>
> Diane: My concern with so much of what you talk about in earth disasters, mass deaths, in an earth change setting, is not the deaths themselves. It's the suffering and it's what the people left behind have to deal with.

Edgar: Exactly. There is no pain in death. The pain is in preparation. When there is enlightened understanding, this pain is greatly lessened. There is a certain sadness of separation, but we often bring the assurance, the reminder, that the separation is neither permanent nor complete.[31]

Edgar says that we have lived many times before and incarnated on many worlds. Earth is not the only sentient planet, and earth herself was colonized from the stars. The material here is an expansion and further verification of earth's extraterrestrial origins. Mary Summer Rain, in the work of the last chapter, states that this information will be general knowledge very soon. For now, it recurs again and again, and is still beautiful and mind-blowing. Read the full transcript of this channeling session in the Appendix, but here is a beginning:

We vision, we remember, there were twelve source planets which made physical colonization upon this one, by thought form manifestation and then moving into reincarnation cycle. In addition to these twelve that made physical colonization, there were many, many thousand that have contributed energy but did not make physical colonization. There are ways in which the energy comes in without needing the physical vehicle.

Just as the United States of America is a melting pot of many cultural traditions, so the earth planet is a melting pot of the universe, of many traditions from many star systems. It is very much a mixed bag of energies. You can delineate and define some of the source planets endlessly, for there are an endless number. But the twelve are the ones we focus on, the foundational ones.[32]

Most of the twelve are planets of the Pleiades system, and two are from the Sirius system — the same ones described by the Dogon people are the ones described here as the foundations of Egyptian culture. Lyra is another star system named, and the people of Lyra were the physical models of the earth body, though Lyrans are larger.

You see, each star energy intermixed eventually with earth energy, for there was, oh, the creatures were prepared to receive the implanting of the spirit. And after many relations, the last to be implanted with the spirit was the prehuman earth plane creature, prepared for such implantation by raising the vibrational frequency. Then the spirit was implanted. In Judeo-Christian tradition, this is allegorically expressed as, "God formed them from the dry land with his own hands, and breathed the breath of life into them." Breathing the breath of life, the spirit, into them — that is what we call the implantation of the spirit. This was done by Yra-AA I colonists, Aztlan (Atlantis) being the first colony.

Diane: Not Mu?

Edgar: That was near-simultaneous. Then after a time of further raising the vibrational frequency of the prehuman and now newly-human creatures over a long period of time, this one now imbued with imagination, free will, choice and so forth, the newly human creature became suitable for interbreeding with star energies. Everyone incarnate on the planet at this time in physical body is the child of Father Sky and Mother Earth. There are sky energies, star energies, and earth energies in you and in everyone. Often the very mixed sky energies (are) from many source planets. [33]

I asked what was the purpose in the creation of earth:

This was designated as a third dimensional or experiential planet. It was a school where people could come and learn, make better choices than they did last time. It was an opportunity. This planet was, from its beginning, a learning ground. There were high entities, positive entities, and there were some not so high, and there were many in between who came here.

The Ra energies complex — Ra, Brahma, Rama, Yahweh, Allah, Ja, Michael, Raphael, Azreal and certain others — the Ra energies complex was and is the group, the energy-bearing supervisory responsible for this planet.

Diane: What do we call those in female?

Edgar: Technically, all these are male and female. However, the progenitor — if you say the Goddess, it is the same thing — Gaea, Ouiramaya, Isis (who was equal to or stronger than Osiris) — these are female manifestations of that generic non-gender energy. Although Allah has been pictured as a masculine god, Allah is as much female as male. [34]

Ouiramaya is also known as Maya (or Maia), and was the first archetype of the Great Mother.

One of the learnings of earth is to bring balance to the power dynamic of the sexes. Where there were matriarchal cultures on earth, Edgar admits, men were not abused as women are in patriarchy. And as he puts it: "Where there has been such male domination for so long, obviously there must be a balance problem." [35] Humans were initially androgynous, and there are planets where three in union are required to reproduce. In one early earth colony, male domination entered this way:

In the Peruvian-Atlantean connection, the Be-ings were Lyra and Yra-AA origin. They were physically narrow-shouldered, narrow hips, long arms and legs, very long slender fingers, feet and faces. And they were androgynous. They are and were beautiful, and they came to this planet. And when the interbreeding began, because of the narrowness of the hips of this androgynous Be-ing, when they interbred with human energy, it was male star energy to female human energy, because the star Be-ing could not safely bear the earth child.

After many centuries, we looked around and realized that because of this physical accommodation to body structure, whereby it was predominantly male sky energy to female earth energy,....that somehow the spirituality aspect was not as strongly implanted in the females as in the males. And they had more sky energy in the males, they were dominant. Not cruelty, simple dominance. And at that time...we recognized that in order to bring spiritual reality at comparable strength to the female line, some would have to go directly into the earth bodies. And there were those who chose to do this,...

And the first of these was called Ouiramaya, remembered as Maya, the Great

Mother. And it was through these re-spiritualized females that female spirituality was restored in comparable degree to (that of) the male. Yet there had been time lapse, and there was much work to overcome male dominance. This was particularly true in some parts of the world, not all.[35]

Today we have virtually the opposite, with female energy being the spiritualized essence, but women still being dominated. It is up to women of today to take back our power, and hope that the males of our culture will accept a "harm none" ethic again.

What the fourth dimension — the New Age, the Age of Aquarius, the Age of Women — will be like is discussed in the next chapter. There is more from Laurel Steinhice, Tanith and others. Earth change predictions have altered radically in the last fifteen or twenty years, and will continue to change as women gain consequence and take Goddess-within power in the future growth of the planet. From the material of this chapter, it seems very clear that women are the hope of earth's survival and the thought power creators of the new world. What we do has meaning now and in the earth to come. What will happen in the coming transition from patriarchy to a matriarchal "harm none" age also heavily depends on us. The change will be "orderly or disorderly" according to our work.

Notes

1. Tanith, *Reading, May 7, 1990,* unpublished.
2. Marion Webb-Former, *On Earth Changes,* unpublished.
3. *Ibid.*
4. *Ibid.*
5. Tanith, *Reading, May 7, 1990,* pp. 2-3.
6. *Ibid.,* pp. 3-4.
7. *Ibid.,* pp. 4-5.
8. *Ibid.,* p. 5.
9. Tanith, *Reading, May 19, 1990,* unpublished.
10. *Ibid.,* pp. 2-3.
11. *Ibid.,* p. 1.
12. *Ibid.,* p. 3.
13. Mari Aleva, *Earth Changes with Mari Channeling.* See Appendix, p.186.
14. *Ibid.,* p. 187.
15. *Ibid.,* p. 187.
16. *Ibid.,* p. 188.
17. *Ibid.,* pp. 189–190.
18. *Ibid.,* pp. 191–192.
19. Laurel Steinhice, *Earth Changes Channeling,* June 3, 1990. See Appendix, p. 194.
20. *Ibid.*
21. *Ibid.,* pp. 195–196.
22. *Ibid.,* pp. 196–197.
23. *Ibid.,* p. 204.

24. *Ibid.*, p. 204.
25. *Ibid.*, pp. 205–206.
26. *Ibid.*, pp. 205–206.
27. *Ibid.*, pp. 207–208.
28. *Ibid.*, p. 208.
29. *Ibid.*, pp. 203–204.
30. *Ibid.*, pp. 197–198.
31. *Ibid.*, p. 207.
32. *Ibid.*, p 198.
33. *Ibid.*, pp. 199–200.
34. *Ibid.*, p. 200.
35. *Ibid.*, pp. 201–202.

The Future

The Age of Women
Future Life Progressions

The choice is each person's...to use personal and group power as it can be used for the advancement of humanity or to use one's Goddess-given potential for the destruction...and the disruption of the Earth. It is the responsibility of all who are awake to make this tenuous balance known, that none may later say they did not understand all that was being held in the balance.

Meredith Lady Young[1]

The whole point of the earth changes lies in what will come after them. For the first time in thousands of years, perhaps even since Mu was destroyed in 50,000 BCE, women will inherit the remaking of the world. Patriarchy has failed the needs of people, animals and the planet, and we are entering a women's New Age. It will be an empty slate for women to write upon, for women to correct the old wrongs by having learned from the past. Even if only a fraction of the earth change predictions manifest, the job will be difficult.

That women are capable of the task is unquestioned; we are training for it now. It has been women who created virtually all of the "harm none" movements of our time. Women were the leaders of the 1950s Ban the Bomb movement, the Black Civil Rights and Vietnam anti-war movements of the 1960s, the women's movement of the 1970s. Women are the moving force behind the current anti-nuclear Peace Movement, the Anti-apartheid movement, the Environmental Movement, and the New Age, wiccan and Women's Spirituality movements. It is (and always has been) women who have said "no more" to such male power games as war, racism, discrimination, nuclear weapons and power plants, and toxic wastes. In fighting for our own rights, women also fight for the freedom and well-being of all.

In the political activism and the psychic atmosphere of these movements, women train for a new world. By conscious awareness of what is wrong in the world, what needs to be different, and then by acting to create the changes and right the wrongs, women choose what the world could be. (Focusing only on the wrongs adds power to them.) In their choosing, women create the thought form that brings that world into Be-ing. Through awareness of a wrong and work to change it, women create the thought form of that wrong erased. By visioning the change and putting that vision into effect, a new earth is born, as step-by-step all the wrongs of the patriarchal world are revised. Women working inwardly, in meditation, ritual and psychic space to visualize a different world, create the thought form as well. Activism and spirituality together change the planet.

Every time a woman joins a protest march, allows herself to be arrested in a nonviolent action, or does a demonstration at a Peace Camp or City Hall, she is training herself to create a new world — and creating it by her action. Every time a woman in the sacred circle invokes the Goddess into herself on the Full Moon, or visualizes a world of peace, health and freedom, she helps it to manifest, along with training herself to create it. Every time an abused woman leaves her abuser and says "no more," she is creating a better world and training herself for its leadership. Every time a woman learns about the environment and teaches her child or her partner to help conserve it, she is creating the new world and training herself as well. The energy for change, and for women's leadership in it, is in these things.

We are all in training right how. With each step in advancing women's awareness of what the wrongs are and what needs to be different, our thought power is creating the new earth, New Age, the Age of Women. By the time of the earth changes, many more women will have awakened and will have become conscious of the wrongs of the planet and aware of how it could be. As women learn more and more fully to take consequence in the world and in their personal lives, the New Age awareness enters into their actions. Women create the new world, and in the creating they learn how to do so.

A seventy-eight-year-old woman in the laundromat told me that the world is coming to an end, and that she feels sorry for what she is leaving her daughters and granddaughters. She has eight great-great-grandchildren, she said, and she is sorry for the world they have to grow up and live in. She can see the old order falling apart, the violence on the news that upsets her, but cannot visualize anything different for the future. The past was better, she said, but the past is over. She said she was glad she would die soon. Some newer women, who perhaps have less memory of a better past, see a positive future beyond today. Aware of the wrongs and of the collapse of the old culture, they look ahead to what will be, what they can create to make a better life for their children and grandchildren. It is up to our generation, and our daughters' and granddaughters' to make a new world. They will birth the newborn phoenix from out of the ashes of the dying old.

Helen Wambach (1925-1985) was a clinical psychologist known for her past-life and between-lives regression work. Her books on these subjects, *Life Before Life* (Bantam Books, 1979) and *Reliving Past Lives* (Harper and Row, 1978), are mainstream classics on these subjects. She used statistical analysis and in-depth sampling of large numbers of hypnosis subjects to discover much about the between life state and the continuity of the soul. Late in her own lifetime she began leading her workshop students into future life progressions, taking them in meditation to lifetimes beyond their current ones or to dates in the futures of their current lives. What she found in their accounts disturbed and frightened her, as she received from her subjects a composite portrait of the earth changes as they were manifesting at that time (1980-1983).

Her subjects, one group taken year-by-year into the coming two decades, reported conditions familiar to readers of this book. They describe severe weather patterns and devastating high winds and storms, with food shortages and inflated food prices. They saw financial and credit problems, and problems with the stock market and bank failures. There would be a scarcity of money and difficulties with invalidated credit cards. Gasoline rationing was reported in the rural areas, to which a number of these presently urban dwellers had moved. There were reports of major earthquakes and volcanic activity, and the sinking of land. Political tensions continued, with the Middle East an ongoing hotspot, and a fifth Arab-Israeli war. One subject reported dying in a nuclear explosion that seemed to be in peacetime, in 1999 in Europe.

These results were summarized in Chet Snow's continuation of Helen Wambach's work, *Mass Dreams of the Future* (McGraw-Hill, 1989). Here, six subjects were taken through in-depth, year-by-year progressions:

> Finally, there seems to be a series of these above-mentioned trends coming together and causing widespread destruction, followed by the subjects reporting themselves released from their physical bodies and unpleasant surroundings. It appears as if there will be a combination of natural and man-made disasters that together wipe out large numbers of people in relatively short order. The six subjects don't agree on the exact date of this general catastrophe but it seems to be sometime between now and the late 1990s.[2]

Note the date of this research, from 1980-1983, the time period when predictions were the most harsh. It would be interesting to progress these same subjects to the same timeframes now. When Helen Wambach received these results of a severely depopulated planet after this century, she decided to quit doing future progressions — but her subjects encouraged her to continue.

> Oddly enough my subjects have been far less uneasy about these results. Perhaps that is partly because all of them report such wonderful feelings of release and joy on leaving their physical surroundings.[3]

With her own health failing, Wambach allied with Chet Snow to continue her research into future lives, and she and her helpers hypnotized 2,500 subjects, 65-75% female. Of these 2,500, only 5.5% were able to describe a future life in the twenty-second century, and 11% described future lives around or after 2300 CE. Her helpers in other parts of the country than California obtained virtually the same statistics. The implications here are that of people alive now, less than 20% will reincarnate on earth in the next four hundred years.[4] These findings verify Laurel Steinhice's channeling of the last chapter, and should not be frightening to those who have read this far.

The research took the subjects into future lives, those who could report future lives, in two categories: 2100-2200 CE, and 2300 CE and beyond. A picture emerges of life in each of these timeframes, although a sketchy one. In the after earth changes period from 2100-2200 CE, there is far less population and four lifestyles are described. These are: small New Age-type communities, space stations, underground hi-tech environments, and rural survival villages. In 2300 CE and after, the lifestyles described are: artificial space environments, other planets, harmonious spiritual communities on earth, artificial hi-tech domed cities on earth, and rural low-tech earth villages. Again, the future lives were based on conditions as they were in 1980-1985, and those conditions have been evolving rapidly. If the same hypnoses were done with the same subjects today, would their future lives also be different?

Only about 5% of workshop subjects experienced a future life in 2100-2200 CE. They reincarnated half as female and half as male (though most workshop participants were female), and a few described themselves as androgynous. They defined their world as negative for emotional/physical well-being. When asked about their deaths in this future life, a few said they did not die at all, and some said they chose to die when their work was finished. One said death came of old age — at 152 — but their average age of death was just less than the world average today of 64.6 years.[5] There seemed to be more control and conscious choice about death than there is today.

Of 133 reports in one group, thirty-five described living in a spaceship, space station or space colony of earth; two lived in stations or colonies of Venus or Mars. They described their clothing as jumpsuits, uniforms or loose tunics, and their food as artificial and bland. Most ate communally or with their families in communal dining halls, some with only their families, and some ate alone. Living quarters were small and austere, not homelike. Some said their spaceships contained extraterrestrials as crew, and they used a universal credit system for money. One subject said she "could materialize creative thoughts," and one woman wrote the following:

> We were part of a communication system spanning around planet Earth and connected up to other galaxies. Our task is to monitor groups to leave Earth during the shift of the axis....I had many successful trips to Earth collecting human beings. I had completed my life's task.[6]

Is it possible that the axis shift will be delayed this long? Or that there could be a second one? Though most of the subjects were white and female, a number described their future lives as male and of other races. Accidents and intrigues seemed to feature causes of death in space.

Unlike the space-dwellers, the women of 2100-2200 CE who lived in New Age-type communities described their lives as satisfying and positive. Twenty-four of the group of 133 experienced this lifestyle in the progressions, and all lived in natural environments of mountains, seaside or desert. The climates were temperate and the air clean, and they described their surroundings as lush. Buildings were natural — of glass, marble or limestone, and many contained greenhouses or gardens.

Over half of the subjects reported teaching and spiritual roles that placed them in these communities for a reason:

> I was in the mountains — windy and cool. It was summer; many trees and heavy foliage. There were steep valleys and rocks with small flat areas. It was a school for children in a convent-like complex; wing-shaped modern buildings jutting out of the mountainside...I was a teacher in the "school of the healing sisters." [7]

When asked to describe an experience with a "bright, shining light," some described the light as a spaceship bringing them either supplies or evolved teachers from other galaxies. Spirituality was a factor in every aspect of this lifestyle, with subjects using such phrases as "enlightenment," "You are light," "wholeness," and "an overpowering sense of love and confidence." [8]

People in this setting wore loose-fitting tunics or robes and sandals. They ate natural foods of predominantly fruits and vegetables, in family settings. A few described dining communally, and some described grocery stores and shopping malls like today's. The people lived longer than the spaceship group, an average lifespan of 92.4 years; death was calm and deliberate. Their communities had a variety of worldwide locations and were sparsely populated. There was the impression of vast earth areas that were uninhabited or unihabitable, and these communities were bright exceptions to the norm. Crystals were used to cleanse the environments, and telepathy seemed to be a general form of communication. This was the most positive of the settings described for the time.

The hi-tech group was the largest of the four types of environments described in 2100-2200 CE, with forty-one subjects (out of 133) describing it. It was not an optimistic environment, but survival-oriented and described by some subjects as desolate and utilitarian. Homes were not comfortable, and people lived in closed bubbles, never leaving them to go outside. Outside was described as poisonous or noxious, with several subjects' deaths occurring from exposure to the atmosphere. There were lung dis-eases, intestinal dis-eases, and an average lifespan of only sixty years. Domed cities under huge artificial bubbles were sometimes described as being underground, in caverns, or even under the ocean.

These were urban environments, "plastic and utilitarian," [9] with rectangular buildings and little beauty. There were metal alloys and what may have been solar panels on building roofs whose buildings were mostly underground. People could not see beyond the domes and never left them, and if the outside world was mentioned at all, it was described as harsh and barren. Few people described delight in their environments or homes, and several expressed frustration or desolation at a mechanistic existence they were glad to leave at death. Wrote one woman:

> I was happy to leave (at death)....I didn't particularly like the group I lived with; it was for survival's sake only. There was no real communication among us. I was sterile and I loved only my few books such as Shakespeare's plays. I guarded them....The others didn't understand about art and literature. Humanistic feelings were lacking. I knew I'd come back in the 2300s to work for a rebirth of humanism. [10]

This lifestyle was emotionally cold.

People dressed in this time similarly to the spaceship people, in metallic jumpsuits, and cultures seemed to be connected. They ate synthetic foods by most descriptions, and said they had little choice of foods. They ate with families or communally, and were in contact with extraterrestrials who brought them food and supplies in spaceships. There was no cash money, but a credit system. There was some spirituality, but it was not always understood or wanted. Several subjects said that the locations of their cities were in the United States.

The last of the four groups, people living in rural frontier type communities or in urban ruins, included thirty-three subjects. They described a lifestyle of about the late nineteenth century, and lived directly on the land in various earth terrains. Dwellings were handmade log cabins or wood-frame houses, with a few of stone or brick. One was a cave-dweller, and all described their housing as primitive. There was little mention of machinery or technology; travel was on foot or by horse and wagon. Settlements were isolated, but seem to have had general stores, and most described them as positive environments close to nature, but lonely.

People in these communities dressed in white robes like the New Age subjects, or in pants or skirts and shirts. They ate meat, fish and game — not described by other lifestyles at all — all natural foods. They hunted and farmed, and used coins for exchange, rather than credit. They seemed to have the necessities of life and to live very simply and without conveniences or frills. There were no mentions of extraterrestrials (unique to this group) or of spaceships. Some may have been religious communities. America, Canada, Africa, Asia and Antarctica were the locations given.

The urban survivors in this group lived in the ruins of cities "seemingly a generation or two after some catastrophe."[11] These were stark and desolate lifetimes, the most negative of the groupings. Sickness is described, as well as rubble, deprivation and violence. Only one subject, however, of all the participants described death by radiation.

Possibilities for these four types of living seem real by today's measure. Space has been hailed as the wave of the future, and earth's participation in intergalactic travel is only in its infancy. Considering that people came from the stars to begin with, it seems possible that space environments and travel would become more a part of future life than they are now. After the earth changes, earth will no longer need to be quarantined for its violence. The hi-tech domed environment appears to be a real possibility, also, an extension of the current trends of patriarchy and technology. If today's practices of destroying the air and environment continue, living inside an artificial atmosphere could become a clear necessity. If today's patriarchal trends toward denying life in favor of cold, mechanistic technology continue, domed cities would be a possible outcome.

The most positive lifestyles are familiar ones to women of today. The New Age communities have their precursors already in places like Scotland's Findhorn or Twin Oaks Community in Virginia. Women's intentional communities have been a facet of feminism for the past twenty years as well. Given that these are spirituality communities, a number of Women's Spirituality groups have projections for starting just such living centers, where women can live and learn together in a Goddess setting. That some of them could become schools for New Age skills — healing, psychic development, ritual training, etc., seems logical and right. Such schools are already operating in some places, on short term, as in the Priestess Training sessions at Circle Farm that run for a week once a year.

The rural communities are also familiar to women, and are a part of the intentional community movement. Women in groups in several countries live on farmland together, growing their own food and creating a low-tech lifestyle. Several women novelists writing about their visions of a female future describe such land projects; more will be covered on them in a later chapter. The women's music festivals are short-term models of this type of working community. Both the rural communities and the New Age ones are described with great positivity as good, healthy lives. If the earth change predictions of cities becoming too dangerous to survive in happen, as is proving true at this time, more and more people with awareness will be moving to the land and attempting self-sufficiency. According to Helen Wambach and Chet Snow's hypnosis subjects, it's a positive lifestyle and choice. Some rural farming communities will also be New Age ones.

Life from 2300 CE and beyond was described by double the number of people who had incarnations in 2100-2200, about 11%. There were 270 in the study, as compared to 133 in the earlier time, and 80% of these participants were women. Again the gender ratio in the future time becomes about half male and half female, with a few androgynous. These people had longer lifespans, averaging 76.7 years (if twenty-eight who died space deaths are subtracted). A few people lived extraordinarily long lives (one reported 400 years), and some reported that they didn't die at all — they transmuted their bodies into other energy forms. Most deaths were chosen ones. Space dwellers were more subject to accidents and violent deaths, and had shorter lifespans; dome dwellers had more violence than the rural or New Age communities. No violent deaths were reported at all in New Age communities or among those living on other planets.[12] Except that space and other planets were emphasized in ways they were not in 2100-2200 CE, the categories are much the same as in earlier timeframes.

Almost half of the 2300-2500 CE progressions occurred in space or on other planets outside the solar system of earth. The 109 women of these groups described life outside of our star system in a variety of planets and lifestyles. Though frustratingly little detail was given by Chet Snow in his book, their reports were mostly positive, and in general, life in the later time was more positive overall. There was space colonization and interaction with other intelligent planets. Extraterrestrial contact was commonplace, and extraterrestrials were usually seen as teachers with superior knowledge. While planet names were seldom mentioned, the Pleiades was named by at least one subject. In other than space categories, only those in the rural communities had little contact with other planets, and in the earth's solar system there were colonies on the Moon, Mars, Venus, Uranus, Pluto and the asteroids. These were cities under domes, with small populations; there was no mass outward migration to them. On earth colonies, in space or on other planets, populations remained small.

Short, belted tunics seemed to be the most frequent clothing for space dwellers, and food was synthetic for more than half the subjects. Almost half reported natural foods, vegetables, fruits and even meat. Most reported communal dining halls and eating with nonfamily members, even when they were permanent residents of the planet or space station described. Living environments were inside atmospheric domes, with buildings of plastic and glass. Travel was by spacecraft, including some personal ones, and there were moving walkways and teleport chambers.[13] When asked about a bright, shining light, most mentioned spaceship lights as commonplace. Some of the spaceworkers had homes on earth in the domed cities, and the spaceships brought supplies to the domes and provided a means of travel to and from them. Most who experienced an outer space lifetime in 2300 CE or after described this lifetime as positive and peaceful.

One woman, living on another planet in a New Age community, gave this description:

At night the adults to outside and sit under a bright sky to meditate in a circle. Children come, too, but they often fall asleep. The adults are in a state of group consciousness, seeming to draw power from the night sky and the universe. Spoken language is not necessary as psychic communication is quite clear and harmony with nature is acute.[14]

Spirituality came from the stars and it is no surprise to find it there again. This sounds like a peaceful life indeed, based on "harm none" women's values.

A few more than half of the participants (142 of 270), described lifetimes in 2300 CE and beyond on earth, rather than in outer space or on other planets. Like the people of the earlier timeframe, this group lived in New Age communities, hi-tech or hi-tech-evolved domed cities, or in rustic rural villages. Life seemed more stable in this time and a little more populous; there was less loneliness and less description of emotional or physical starkness or sterility. There were more children here, more peace of mind.

The New Age communities of 2300 CE and beyond seem appealing places to live. They were close to nature in temperate climates and described as beautiful and green. As in the earlier group, the communities were in natural settings, in the woods, by the sea or near freshwater lakes or streams. One woman described her life in a community of healers and teachers in the Canadian border area:

> The landscape is woodsy and cool. Beautiful surroundings. There are trees, shrubs, a small waterfall and stream. I see a white library building shaped like this (a cutoff pyramid), as are the surrounding homes...It's welcoming and delightful. [15]

She described herself as female in this life, wearing a soft, floor-length white dress and leather sandals. Her community was communal-based and family-oriented. The buildings were low, white or buff-colored and surrounded by green gardens. People could teleport short distances. It sounds like a women's utopia and is described as a new Atlantis, though these communities were larger and more urban than the earlier ones. Harmony with nature and each other was emphasized and the food was vegetarian.

Despite the positive alternative of these types of communities, the hi-tech domed cities were also still existent, although earth's ecology seemed more stable. The domed cities were still sterile and emotionally unfulfilling, and without contact with nature. The outside environment was hostile and inside the domes were violence and loneliness. No one described this environment as beautiful.

Buildings were of plastic, concrete and glass and fully utilitarian. The people in this environment gave an average age at death of 56.7 years, the lowest of any on-earth group of the time. New Age lifespan average was 99.6 years in 2300-2500 CE. In the domed cities, the cause of death was often violent or from respiratory dis-eases. The women gave various locations for their homes, including on land that was formerly part of Atlantis.

One woman described her domed city in this reincarnation:

> I was alone, looking at a very cold, tall and metallic-type city which you couldn't see into. The buildings were all close together, touching, no space between, no windows and reflective surfaces, a greyish-green color. I was tall and slim, all

encased in a body suit, except for face and hands....There was a huge empty space all around. Nothing green or growing, no trees, no other people, all was encapsulated in the city, I guess. I was not aware of my sex; it didn't seem important. My hair, head was covered with the body suit. [16]

Some of the domed cities were more positive, described as hi-tech evolved. These had similar futuristic surroundings but also a relationship with nature. The dome could be left now, at least temporarily. The hi-tech environment seemed to be moving more toward the New Age communities, with more humanistic values and less sterility and loneliness. There was no violence reported, and no deaths by respiratory failure. People were happier and lived longer (70.9 years average). Spacecraft were mentioned, as well as healing energies and telepathy. There was caring for others. [17]

Apparently the harshness of city life was softening, and spiritual and "harm none" matriarchal values were becoming more widespread. The land was less barren and no longer poisoned around these settlements; there was contact with nature again as the earth regenerated. There were mentions of psychic skills as part of everyday life, telepathy and healing — the spirituality skills of the New Age were mainstream now. Perhaps these evolved cities appeared later than the other high-tech environments. They were certainly more positive as places to reincarnate into. Again, a risen Atlantis was mentioned as the location for some of these cities — it was also mentioned for the New Age communities and the less positive hi-tech ones.

Rural, rustic outposts were still described in the 2300-2500 CE timeframe, and these too had changed and evolved. They remained natural and land-based but became more modern and comfortable. It was still a frontier, however. People were trappers and farmers, and one location was described as Greenland, now a temperate climate being colonized. Population was increased, and there was more evidence of technology — hovercraft along with horse-drawn wagons. There were open-air markets, and spaceships bringing supplies. There was electrical lighting not mentioned before. It was reported as a physically tiring and hard life, but family-centered and warm, and mostly described as positive. There were a few survivor-types, soldiers or scientists exploring old ruins in desolate surroundings, somewhat different from the survivors of 2100-2200 CE. They were there only temporarily.

These future life progressions, done by Helen Wambach and her associates in the 1980s, are intriguing in the least. They give a bare outline of what life in the future after the earth changes will be, but also leave many questions unanswered. What will governments be like? Similar communities are described in widespread geographic locations — is there one world government? What there clearly is *not,* is a New Age nation, a hi-tech nation and a rural nation; all of them exist in all places. Why is there such an attitudinal difference between the New Age-type and rural vs. the hi-tech and space dweller communities, particularly in the earliest timeframe? Why do people live in the less positive environments at all? Do they have choice?

What Laurel Steinhice calls fourth dimensional (and Sheila Petersen-Lowary fifth dimensional) skills seem evident in the after-earth changes future, but are more often part of daily life in the New Age communities than in the other lifestyles. These would include such things as teleportation, telepathy, and use of meditation and healing. Is it only the New Age and other planet dwellers who use these skills? Why are they ignored in the domed cities until the later hi-tech evolved ones? Fear of death seems to be gone in all of the lifestyles, with death described often as a choice and accomplished in communal settings. Again, there is more emphasis of this in the New Age-type communities.

How long will it take for the earth herself to regenerate after the earth changes and the ecological disasters that precede them? How long will it take for civilization to evolve again from survivors and rural frontierspeople (if that was the start) to the larger New Age urban communities or the hi-tech evolved ones? And what is the process? Will these highly divergent lifestyles evolve simultaneously or are they the features of separate time periods, and if so, what evolves first, second, etc.? The New Age communities seem the most positive from beginning to end, and from the information of channelers are what the earth is moving toward in the Age of Women.

Eventually hi-technology and spirituality will merge, as is indicated in the later New Age urban communities and the hi-tech evolved societies. Are the other lifestyles transitional? From 2100 CE on, life seems to become more positive and caring as time progresses. New Age communities start small and later become larger and more urban but always remain close to nature. Hi-tech and space-dweller communities have initially no contact with nature at all, but move more in that direction. From the sterile, cramped space stations people move to other planets, where life is again more humane and natural. Population doubles from 2100 CE to 2500 CE, but is still only a small fraction (less than 20%) of the population on earth today.

The different lifestyles are marked by geographic and racial diversity, but there seems no distinction between the Black, White and Asian cultures. Will people at last be one mixed race, designated only as human, instead of divided into segments? What will happen, if this is so, to the rich cultural diversity of peoples? Or is it because the majority of hypnosis participants were white that this issue was not discussed? Or because it was uninteresting (or unnoticed) by Chet Snow, who interpreted the information and chose the examples to print? Settlements are described frequently as being in Africa, the United States, Greenland and Antarctica, Canada, Australia, South America, Asia and in other locations. Never is there any discussion of an ethnic or racial culture. Nor is it mentioned for other planets — where very little other description is relayed, either.

One woman described her mission in the later timeframe would be to rehumanize the hi-tech cities. From 2100-2200 there seemed little emphasis on aesthetics, art and literature, music or in creating a gentle way of living. From 2300-2500 there is more of this, more humanism, color and culture, with a more positive outlook and way of life even in the domed cities. The 1990s are probably less culturally oriented than, say the eighteenth century: is this lack of culture a trend of the twenty-second century, or is there further reason for it? Is it that life is still survival-based in 2100 CE, with less time, interest and resources than will be available later? This lack of culture seemed less apparent for the New Age communities in either timeframe, perhaps because of their natural settings.

There is a gap in this information, too. The times leading up to and just after the earth changes (before 2100) are not discussed at all. Helen Wambach gave her subjects the two definite choices of dates to experience their future incarnations. She did not ask for or make leeway for reincarnations in the twenty-first century or at other times. Had these been available, information might be clearer as to why the communities developed as they did, and on their divergence in style and attitude. Information is not available as to how a woman chose a New Age community over a domed one, or life on a space station over life on land. Was she born into these communities, did she choose them, how did she make the choice, could she change her mind? Why did the apparently negative lifestyle of the domed cities develop, and why did it develop so negatively? How much of the 2100-2200 CE time period was transitional?

Extraterrestrials and life on other planets were mentioned in all of the lifestyles and considered commonplace to these future inhabitants of earth. The rising of land from Atlantis and its recolonization is also mentioned, particularly in the later time. Various spirituality skills are mentioned. With the attitude of death as a choice and transition, the concept of reincarnation must also be given for these future people. Where the land is clean and unpolluted, New Age and rural communities that live close to nature are possible. Where the land is damaged and the air noxious, domed cities are the choice and probably the necessity. Where the land is clean, natural foods are available; where it is not, food is synthetic and unappetizing. Most of the subjects described communal lifestyles for their future lives.

Where will women of today choose to reincarnate in their own future lives? And will these future lifestyles change in their projection as women's awareness changes, in the way that the earth change happenings themselves are dependent on women's growth? It is possible that the domed cities and negative space environments need not develop, that the New Age and rural and hi-tech evolved communities can emerge early, also determined by women's awareness. Women are already evolving communal living as the wave of the now and future. How much else are we evolving in our changes of today that will affect life in 2100 or 2500 CE? These are all things to think about, as women create the

New Age, the Age of Women, in our growth and changes within.

Sheila Petersen-Lowary, a channeler and author of *The Fifth Dimension: Channels to a New Reality* (Simon and Schuster, 1988) has some ideas of how the changes will come about and what will be important or lost in the transition time. She speaks of evolution in governments and land masses, in individuals and relationships, changing from patriarchy to the New Age. With her guide Theo, she talks about the shaping of the future:

> We speak futuristically for the truth of the moment. Understand that there are no absolutes, for the will...is the creative force, the creator of what is to come about. So when we speak of the future it is truth seen for this particular time. As beings evolve and the direction of their will changes, they can alter the forward movement of events.[18]

By this information, the radically different lifestyles presented in Helen Wambach's work may be choices, alternate realities, possibilities to choose which will manifest. In the next hundred years, say Theo and Petersen-Lowary, the earth will change. Despite the earth change predictions, this change is optimistic:

> The world as you know it will not be. The consciousness of humankind is being raised, the energy is being more refined. There will be a rebirth, a rejuvenation of this planet. Political forms will change. Economic structures will change. The old forms will no longer be. This will happen in the near future...immediately...in your timeframe, during the next five years.[19]

Governments will change and are changing, to be more sensitive to people's needs. These will be world changes, as well as changes within the United States:

> What comes about is a unification, a combining of forces who share a vision, a combination of thought. No one personality will lead, but there will be a restructuring, so that,...a committee will be governing....Much input will come from other groups who do not now have access to political power, from the artistic and scientific communities; for it is these beings of sensitivity who will first acknowledge the needs of the planet and its healing....
>
> The United States government will change the nature of the presidency. There will be more than one personality making decisions. What you call the "Cabinet" will have more power. There will be a group making decisions.[20]

The Cabinet will be elected in this country, and there will be a unification, a committee, governing the world. The feeling in the future life progressions of a single government, of no more nationalistic structures, could evolve in this way. Government will be global, and for the good of *all* the people. Tired of wars, food shortages, lies and poverty, people will unite to force these changes upon government structures. There will be cooperation instead of aggression and repression. Petersen-Lowary says these changes are already beginning (in

1988), as felt in world unrest. There will be the shifting of government structures by 2012-2015 CE. The daily news shows it happening now, in 1990.

There will be more cooperation between governments in the arts and sciences. Communications systems and satellites, space research, computer research will be shared among cooperating governments. There will be a merging of technology and consciousness, medicine and healing. Discovery of life on other planets will be made and made public, beginning earth's contact with extraterrestrials and extraterrestrial cultures, and beginning the future of space stations and routine spaceship travel:

> Yes, there will be the discovery of life on other planets, and there will be further space exploration...before the year 2000. These discoveries will have a great impact. Through these discoveries there will be greater interaction with other nearby planets — especially Mars. But the discovery of intelligent life in other systems will wait until humankind is prepared in their consciousness to accept it. It could have happened long before now, but humankind was not prepared. They are preparing now.[21]

Many women saw their future lives among the stars, and this will begin very soon.

She continues:

> Know that there have always been visitations from other planets, other solar systems. Especially from what is called Pleiades and Sirius. There is much Pleiades energy present now.

> There is a knowing within humankind, a memory brought forth from the past. The information will be imparted into the world and shared. The information is eternal, and humankind is being prepared now for this knowledge. The opening to the universe is part of this.[22]

This is information we have heard again and again, through a variety of sources. The knowledge of human origins will end earth's quarantine from the rest of the universe; we are being prepared for this knowledge which is an important part of the New Age, the Age of Women.

Another important part of New Age awareness is the balancing of the power dynamics of male and female. Sheila Petersen-Lowary sees this as the balancing of the right and left halves of the brain. The intuitive and concrete minds are compared to female and male energies, respectively and every Be-ing contains both energies within them. In the coming New Age, what she calls the fifth dimension, both halves of the brain of the individual, and both genders will learn to respect and understand each other. The male will no longer have license for abuse of women; there will be full equality. The intuitive female mind and the concrete/technical male mind will both be given training and validation in everyone. Once there is equality of these energies within individuals, a balance will be struck between genders.

This manifests in relationships, of course, and both opposite sex and same-sex relationships will change:

> In the 5th Dimension, relationships must be created wherein the trust of one's self blossoms forth. As you develop this trust in the totality of your being, you allow yourself to become vulnerable and to accept one another....In this comes strength — taking on the responsibility for your personal learning and also openly supporting the experience of others. Then the release of judgment comes, yes?

> But first the anger has to come out, and many are angry now. You will see this come to the forefront. There will be upsets, and communication will be angry at times. It is a part of the release of the 5th Dimension energies that are at hand. [23]

How people treat each other in general, and women's release of old pain and anger, is part of the women's New Age. Men will learn to respect women, and relationships including lesbian relationships will be based on a new trust that begins with self-trust. The many women who are coming to terms with and releasing the pain of incest or childhood abuse are part of this process of balancing and releasing. The many relationships coming to an end at this time, even long-term ones both same and opposite sex, are also a part of this. Relationships not built on trust and self-trust are ending. Women are dealing with their anger, anger from within and anger at oppressions without. We are saying "no more" to all the forms of gender oppression that men have placed upon women.

In opposite sex or same sex relationships, the form of commitment between partners will change:

> The commitment will be of the heart, of sharing, of common experience and lifework. Independent individuals will come forth and share their strengths in their relationships. There is no longer only the passive and the dominant roles. That is where manipulation occurs. Relationships in the 5th Dimension will involve the coming together of peers, of those who are equal in strength and certainly within their selves. This is important, not only for individuals, but for the planet. The planet is in need of this balancing created by humans existing in love. [24]

Many women are already learning this form of love, a far different form than our patriarchal upbringings taught us. It is free, healthy and liberating. It is matriarchal and "harm none." These are the relationships that will be able to accomplish long-term and lifelong commitments. They will affect the growth of earth both now and in the Age of Women.

The growth and evolution of intimate relationships will affect the growth and upbringing of children; and children are the New Age in their very Be-ings. When adults learn balance, trust and self-worth in themselves, they will teach that to the children. Women becoming aware in this time are focusing on their own childhood needs that were unmet, their own sufferings as children from abuse, power-over and incest. By knowing and releasing that energy from within themselves, they break the cycle that passes abuse onto coming generations. By freeing themselves, women free their children, and create a different kind of people to make a more loving age.

The children who are entering the earthplane now are coming here for the purpose of creating a new world. There are many advanced souls among them, and many who were leaders in the matriarchal world of Mu. Many of these children are highly psychic, living in a world that their parents must learn about from them. Validating their children's experience, encouraging their psychic gifts and raising them with love is the greatest good any parent can give to the planet.

> There is now an acceptance of being, the honoring of what is the wholeness of the human spirit. The perfection, the divinity, of this being must be communicated. Speak to the children of this divinity and its acceptance within themselves. It is important to nurture these beings in this attitude, to develop a nurturing sense of acceptance and love. These beings will demand it, do you understand?
>
> They will teach their parents more than the parents will teach them. There will be an exchange, yes. But these beings coming forth now have great wisdom that will be remembered. It will be well for the adult to listen to the child.... [25]

Children need to be taught "harm none" values, and the Goddess' rules and redes (in a Goddess context or not). "You are Goddess" is the divinity within. Unlike their parents, these children are to be raised with self-worth and freed from the limitations and pain that their parents and grandparents grew up with under patriarchy. By raising free and whole children, children who have strong self-worth and self-trust, children who are not subjected to the abuses their parents survived, women create the New Age and earth's future. Our children will be free, and they will free the planet. They will be capable of creating the cooperative, communal lifestyles described in the future life progressions to the next centuries.

Helen Wambach's progression work paints pictures of what the New Age, the Age of Women will be, both positive and negative. Remember that women create reality, and that all reality is choice. Sheila Petersen-Lowary presents ideas on how that New Age will evolve, how we will get from here to there. The future is not created overnight, but in steps. Each step is a choice that builds upon the choices before it and determines the choices to come after. Each woman's choice today creates the future. Women, in their thought power, are creating the New Age now. Do we want the future to be the patriarchal domed cities or the matriarchal New Age healing centers? Human reality and consensus will give us what we ask for.

Notes

1. Meredith Lady Young, *Agartha: A Journey to the Stars,* p. 185. I have changed "God" to "Goddess."
2. Chet Snow, *Mass Dreams of the Future* (New York: McGraw-Hill Publishing Co., 1989), p. 45.
3. *Ibid.,* pp. 45-46.
4. *Ibid.,* pp. 36-37.
5. *Ibid.,* p. 144.
6. *Ibid.,* pp. 119-120.
7. *Ibid.,* p. 124.
8. *Ibid.,* p. 126.
9. *Ibid.,* p. 132.
10. *Ibid.,* pp. 133-134.
11. *Ibid.,* p. 146.
12. *Ibid.,* pp. 156-159.
13. *Ibid.,* pp. 165-166.
14. *Ibid.,* p. 178.
15. *Ibid.,* p. 185.
16. *Ibid.,* p. 200.
17. *Ibid.,* pp. 203-205.
18. Sheila Petersen-Lowary, *The Fifth Dimension: Channels to a New Reality* (New York: Simon and Schuster, 1988), p. 63.
19. *Ibid.,* pp. 63-64.
20. *Ibid.,* pp. 74-75.
21. *Ibid.,* p. 82.
22. *Ibid.,* p. 105.
23. *Ibid.,* p. 112.
24. *Ibid.,* p. 115.
25. *Ibid.,* p. 151.

The Age of Women

Thinking it into Creation

Our way out will involve both resistance and renewal: Saying no to what is, so that we can reshape and recreate the world. Our challenge is communal, but to face it we must be empowered as individuals and create structures of support and celebration that can teach us freedom. Creation is the ultimate resistance, the ultimate refusal to accept things as they are. For it is in creation that we encounter mystery....

Starhawk[1]

What will the new world be, the Age of women, New Age, Age of Aquarius? What happens after the earth changes, after the pole shift, after the earth has been purified of pain? When negativity has been cleansed from people — old karma and hurts released, unworkable ways let go of, violence and abuse "shipped out" — what is the society that women will create? Imagine a world where women can go out alone at night without fear. Imagine a world where all races and nations respect and honor each other. Imagine a world where trees and clean water are valued, where animals are treated with dignity and caring. Imagine a world where women and men both know women's consequence and live in "harm none" ways.

Think of the best possible world. Think of a world that is ideal, utopian, everything you could possibly imagine of peace, freedom, beauty and love. Think of what governments would be like if they cared for their people. Think of what people would be like if they cared for each other. Think of what the world would be like if the earth were taken care of. What would the New Age be, the women's New Age, if it were run on "harm none" Goddess/matriarchal principles? Think of the best possible world, and then go ahead and create it.

Women's thought power can create this ideal world, is creating it right now. Think out carefully what the new world will be, visualize it, understand it, see how it can come about — and put it into effect. Think it out carefully, for "what you send out comes back to you" and "be careful what you ask for, you might get it." Realize the implications, who will be affected by any change and in what ways, then create a new world. If it harms or manipulates any, it is good for none; if it helps and frees any, it is good for all. In the cast circle, in

140

meditation and visualization, on the protest lines, in writing and speaking and teaching, in the voter's booth, women create the new world. Like Euronyme and the cosmic egg, all things are contained within the thought. Like Spider Woman, thought and creation are one.

There are many possible realities, and what women think influences the choices of what will manifest. Whatever can be thought of can be made real. Therefore think only in positive ways, create/think the world in ways for the good of all. Think abundance, love and peace for all, security and freedom. In rethinking the world today, women create it for tomorrow.

It is women's changing awareness that is giving earth the opportunity of the earth changes. The changes are a chance for the planet to release and heal the thousands of years of patriarchy and to return to balance. They are a chance for Gaea to regenerate, to be rebirthed. They are a chance for women also to be reborn, to heal their pain and replace it with joy and love. Think of the earth changes as a positive opportunity and they become so. Think of the earth changes as the beginning of the Age of Women, of a new matriarchy, and help that New Age to gestate towards its birth. She is already in the womb, already entering labor, already crowning. She will soon be flying free.

Mary Summer Rain, in *Phoenix Rising,* describes the earth after the changes. She sees a reborn planet, alive with hope and beauty:

> I peered down and saw a world at peace. The beautiful orb slowly rotated beneath a veil of whispy white clouds. It was so magnificently restful. I wanted to cry, but what really brought on my tears was the rosy aura that pulsated from the earth — it was the aura of absolute love.[2]

This is a far different planet from her earlier aerial view of the redesigned world, and a far different earth from today's.

Government will be for all people, and all races and nations will be equal. It is the awareness of people that will stop the natural and man-made disasters, and the awareness of people that will create new and peaceful ways. There will be a change in basic values, from the patriarchy that didn't work to a "harm none" matriarchal way that does. Native Americans, and others who have "walked softly on the Earth Mother," will be listened to and respected as those who know how to live in the paths that earth is seeking. This includes women, who have recreated the skills of consequence, "harming none," and living in equality and peace. Women of all races and cultures have done this from the beginning of incarnation on the earthplane.

In her astral journey to the future, Summer saw herds of cattle grazing free, all ages together and in families. None were to be eaten. She saw acres and acres of food crops — wheat, corn and alfalfa — growing tall and strong to feed the people. There were houses being built, with manual rather than power tools. The houses were circular and partly underground, and fitted for solar power.

She describes a "church" or meditation center:

> It was outside in an open area. People were seated on the ground. They had their legs crossed. Suddenly, they rose in unison and raised their arms to the blue sky and gave thanks. They thanked God(dess) for the rich earth. They gave thanks for their flowing waters and the warm sun. And they thanked God(dess) for each other....There didn't appear to be a leader who conducted the ceremonies. They acted as one unit. [3]

Life in the Age of Women will be close to the earth and respecting her, living on her bounty without draining her resources or enslaving other Be-ings. Animals will be free, as will people. The earth will nourish and nurture as she always has, but this time without men raping her body or exploiting her lives. There will be no more pollutants; building and heating will happen in safer, non-destroying ways. All people will "walk softly on the Earth Mother," living in respect for each other and the planet.

Says Mary Summer Rain:

> This was the ultimate end for the changes. This is why they began in the first place. Now, all races will be equal. All...will live and work and play as a total unit of humanity. The barriers will be torn down. But what will be most beautiful, will be the way in which people live. No-Eyes showed me a most unique new civilization whereby technology developed innovative energy sources. The problem of gravity was conquered by a method which enabled people to reverse and control the magnetic polarity of the earth. This one discovery has endless ramifications.

> All the lower life forms will be allowed to roam untouched. People will have learned to eat according to the Earthway. They will have devised new construction techniques and structures. They will cultivate high-yield protein crops and will have halted world hunger. They will join in a universal worship and recognize their planet as a living entity that freely supplies all their needs. They will be one in...humanity. They will be one with all of life as they recognize the peace of living the Way of the Indian Nation — the Earthway. [4]

She describes the sound of Gaea's heartbeat, heard by aware people from deep within the earth. Those who hear it do not speak of it, they are waiting for the others to awaken. Those who hear it come from all races, all parts of the world — and most of them are women. It is the women who hear Gaea's heartbeat that are creating the new and better world.

Marion Webb-Former, writing on June 23, 1990, describes the future. She warns, however, that the future is still changing and predictions cannot remain constant because of free will and choice. Her entity Moon places the re-emergence in Africa:

> Africa is still the land where the New Beginning will truly take hold. Many will flock to her continent for what they perceive as their own personal reasons, but once again the group soul mind will be guiding this decision....

What would you have me tell you of the future?...Many, many facet souls...will have removed themselves by succumbing to the catastrophic earth changes and the AIDS epidemic. Those who remain will be far from perfect....Yet others, who have found the true path, will be awakening to the enlightenment of ALL THAT IS. Physical life for many will be difficult as it was for those who experienced "frontier days"....For those who will have experienced the light side of the New Beginning, it will be their faith in their spiritual being which will give them tenacity. Within this group will be many watchers and creators who have "walked-in" during the "latter days" before the axis shift. For your understanding, consider this biblical term to mean the second half of the twentieth century.[5]

Moon describes vastly altered living conditions on an earth scarred by pollution. There will be many species lost to extinction but people will survive and rebuild. Souls who once lived on Atlantis and Mu will reincarnate and establish a New Atlantis in Africa:

The technology of crystals, gemstones, pyramids, electromagnetism, and energy vibrations will be used to advance this nation to even greater heights than were achieved by the first Atlantis.[6]

These people will merge culturally with those living on the risen portion of Mu. The combination of New Atlantean technology with new Mukulian spirituality will herald thousands of years of peace. Marion and Moon report future incarnations of earth Be-ings on other planets and colonization of the earth's moon.

In three readings, Tanith writes extensively about the future of earth after the earth changes are completed. She describes a world that is both different and familiar, with people and their concerns — particularly women's concerns — very much the same in future times. She describes a world of harmony, of living with the earth, and a world of quiet joy:

There will be small towns, more like networks that work together to raise children, raise food, make clothes and shelter. They will again celebrate the ancient turnings of the seasons and changes of the year. They will grow old in harmony with the earth, fire, water and even the air. The air will clear, and other creatures will again outnumber humankind. We will not lose our writings, or our learnings, but finally understand how they fit together and intertwine with the needs of the earth. We will develop our minds in more powerful ways, and rejoice in the knowledge that was always hidden deep within us. We will finally learn from the creatures that surround us, for we shall finally be still and listen to what is said.[7]

Will everyone get along and be happy? Of course not. People are people, but many will. The changes are what is best for the earth as a whole, which means the butterflies count as much as we do. The opinion of the otter and beaver and turtle hold as much weight as our own. The changes are best for all, but will not make all people happy.[8]

In her May 19, 1990 reading, Tanith describes a New Age that is truly a women's New Age. Throughout her work, she emphasizes women's role in creating a new world. The values of the future will truly be women's matriarchal Goddess values:

> Women will gain in importance, strength, self-esteem and awareness. They will take over much of the leadership roles as teachers and healers and guiding lights. Education and wellness will take precedence over money and power. Personal power will be defined in terms of enlightenment, awareness, creativity, and personal growth....

> There will be less people on this small planet, and the human species will not take so much space or energy or resources from the Mother Earth. This will allow the planet to heal herself and be renewed....[9]

At Summer Solstice, 1990, Tanith taped a crystal ball reading about the world to come. What is familiar and what is different are highlighted in this extensive description of the women's New Age. Her reading discusses everything from raising children to racial equality, from women's status and compassion for others to the Goddess' Crone aspect. She talks about animals and other planets, technological breakthroughs and women's psychic networks. It will be a very different world, but not an alien or estranging one. The full transcript of the reading is in the Appendix, but following are some of the highlights.

After the process of earth change ends, many will be happy with the new world, but some will find the new ways not to their liking. They will miss things no longer available and be unable to adapt. There will be what Tanith describes as "a second cleaning," where those unable to change and those still immersed in violence will be weeded out. This will happen nonviolently.

> There's going to be a second weeding, and that second clearing is the ones who thought they could do it and now get there and realize they want the old life. They're the ones who still perpetuate the violence and the crime and the...

> Diane: How will that second clearing out happen?

> Tanith: The ones that like the violence will sort of segregate together and the others won't. And the ones who are not with them will be the ones who survive, the ones whose neighborhoods can survive, the ones where disease does not — it's almost like the other ones forget. They just go all the way back; they slide all the way back. They forget all the things that kept them healthy in living through the times and then they slide all the way back. See and it isn't violent destruction....[10]

I asked her what would be gone, what parts of modern life will be lost to us after the earth changes:

> A lot of the things that are real luxuries that are considered essential. And I don't think they'll be totally gone. I don't see that we won't have electricity, I don't see it the dark ages, but I see that conservation is so incredibly important that we don't simply waste it. We don't waste the electricity, and there'll be other things that

we'll be doing so that, uh...the movie industry won't be as big. We won't be making the movies so we won't have a new slew of movies every summer and movie houses to go to. There still might be movie houses but it's not going to be the brand new movie every single week coming out hi-tech, because the tech will be going for something else. Some things in the home — home will be pretty much okay — it's not the dark ages. It's not like we're going to lose washing machines and dryers, but there's going to be less use of VCRs cause there'll be less tapes to play in your VCR, and the TV won't be the main source of entertainment anymore. [11]

Creativity will be valued over technology, the opposite of today. Storytellers might be more the entertainment mode than TV, and creativity in children will be encouraged. Where there is technology, it will be technology that *does* something or fixes something, rather than technology for its own sake or mindless gadgetry. Fossil fuel to run the motors will be less accessible, and creativity to develop new methods will be welcomed and important.

Some things will be gone because they won't be useful. Gas is going to be very very expensive, gasoline. And so you might still own your car and there might be gas stations, but you might have to be a millionaire to drive it, because it won't be accessible like it's accessible now.

Diane: Will there be any transportation that's accessible?

Tanith: Yeah, I think that there's going to be solar cars and wind cars, and I don't know about planes — something about planes, there's something different about planes. Almost like we have this leap in technology that occurs in like ten years or twenty years. This leap will take us into the Star Trek generation practically, this leap will take us to that different form of energy that allows us to commute in a form that's something else and non....

Diane: Antigravity?

Tanith: Yeah, yeah....Even some of the car vehicles will be doing that, too. Just regular vehicles, motorcycles, things like that will be not with wheels, they'll be antigravity. They'll be skimmers. We'll just take this leap in technology, and that kind of creativity will be really really appreciated, really valued. [12]

From what is lost in the way of fossil fuels and modern technology, a new form of power will develop, or several forms of power. These forms will be ecologically sound, and the creativity of their inventors will be valued. Today, children's creativity is stifled and adults with a new idea have a very hard time marketing it. This will be a major change in the women's New Age.

Communities will change will focus on local networks, rather than on national governments. Government at the federal level will be disorganized at best, and local coalitions will take over. I asked, "Will government have more compassion in how people are living?" Here is Tanith's reply:

People will have more compassion about how people are living, and government will just be ignored if it doesn't. People will just do it themselves. There'll be more backyard coalitions that form to plant backyard gardens to feed all the neighbors and make sure that every old lady in the place has a house and has food, and no one will ask the government. It's like, can you get government aid, well no, but why would we need it anyway? That's not what we need, we need neighborhood aid.

People are going to start taking care of each other because they'll understand that it benefits each other. A lot of people who haven't had jobs for ages may find that they are incredibly needed in that coalition pulling together. Some of them will feel like we've really declined because now we're back to sort of just living. It's not a survival issue, it won't be like we're all going to die off if we don't do this, but just that everybody's taken a step back and said, well let's enjoy life. So a lot of the intense competition in businesses, in money, in government and technology won't be there because no one'll really care. [13]

Women will have a big part in the change from centralized to localized government, and it will be a change based on women's "harm none" values.

The country's going to have more autonomous states and towns and city-states and there'll be basically that cover over it that says, okay we are all Americans and we will all have that in common, and we'd all fight for something if it came to it but it's not going to come to it. And basically it's going to be women saying, we need to feed our kids. I think it's really going to come that, it's going to come to a food issue. We need to feed our kids. We aren't interested in the rest of it, we need to feed our kids.

And much of what the United States government does will just be ignored because it's international, and the international's going to stop a lot because people are going to say, heal here first before you heal out. And a lot of the countries we're giving money to may not be there anymore to give money to, and we won't have any money to give them anyway. So they may make policies about how they talk to other countries but it won't really matter. So at most the government will still come back down to city. Most of the government now that affects people is city-run, state-run, and it's still going to keep doing that falling down. [14]

There will still be a Congress and President, but the government will have far less power. Women will simply ignore them, listen to them less, and live their lives. By withdrawing their support and participation, women will effect change.

We're going to see a change in that women just stop listening, stop allowing it, cause women have participated, they have allowed this. And they will just simply stop allowing it. It isn't violent. It isn't a violent change, it's just, I don't have to do that. And no you can't have my money to put into your bombs, I'm not interested, forget it. And by that point there won't be any way that the government can enforce the collecting of taxes because they'll be breaking down. [15]

The war machine will break down, too. Tanith sees a different role for the Pentagon.

We need the changes to go through more, so there is no Pentagon to destroy anything good that comes, and try to dissect it and figure out what it is.

Diane: Will that happen? That there'll be no Pentagon?

Tanith: Yeah. It'll be there, it'll be a big building with a bunch of good old boys who think they're important and no money to support it. They'll have some things. They'll have a National Guard to come help the people after a flood; they'll have a navy to watch the shores. More of their job will be pollution watch, animal watch — it'll be a watch. It'll be there because the men who have moved up in those ranks need somewhere to go, and so we'll give them a ship and tell them to patrol the shores and make sure we're safe. They'll all feel very proud of themselves, and they'll leave us alone. It's like giving a child a toy and saying, please play with this and leave me alone, and so we'll do that.[16]

In a world based on women's "harm none" values, children will be raised differently from how they are now. The new culture will be cooperative and local, with an acceptance and validation of the individual. There will be a downplay on technology and a focus on human needs and resources.

It's going to be smaller groups, and I think what we're going to find is that the schools will be run by parents. There will be schools, because schools are just so efficient, and because schools allow so many other adults freedom. But the teachers will be parents of the students, and what we will do is take the people that we highly value. Older people, too. We're going to take old people and make them teachers, when they retire and don't do other stuff anymore. We make them teachers because they're good for the kids and safe with the kids, and because they have so much to share.[17]

Children will no longer be punished for their creativity or for their psychic skills. Tanith feels that psychic ability has been somewhat dying out because it has been repressed in children. She feels that it will return with creativity:

I think we're going to be finding the strain of it is just a little stronger. It's been dying out somewhat in children in the same way that creativity has, and trained away, and there are many women who have the power not bearing children. But what you'll find is as we let that creativity come back and that acceptance is just there, it's not taught away, then that will come back.[18]

The validation of children will be the validation of all races of children. But while racism will end, true equality is still far away. I asked what this would be like.

Well, you know, I'm not sure. People are people. Black horses don't like white horses. I think what you'll find is a greater equality in the numbers of races worldwide, but possibly not the mix. It may not be totally mixed per country. And it may not seem fair,...but people are still going to remain people in many ways. They may be more tolerant, and that will continue to grow. Equality is a weird word because we're not going to see world equality right away, where everyone has the same standard of living. Not right away.

Diane: But will we stop seeing the kind of thing where we're putting you down just because you're Black, or you can't have this job just because you're Black, or you're not good enough just because you're whatever color?

Tanith: No, that'll be pretty past. It will be one of those things where if it happens it will be recognized fairly quickly and the community as a whole will not allow it....And the person who is like that will find they are ostracized. They will have the pressure on them to match the whole, and the whole will be one that is a lot more open.[19]

With a greater respect for all of life, there will also be a greater respect and understanding between people and animals. Tanith's dog Onyx told us repeatedly during the reading that "Dogs are important," and will be much more important in the women's New Age. Tanith mentioned dolphins, whales and global warming. The dolphins are our teachers.

People will realize that, yes dolphins do have great gifts, in fact dolphins are more intelligent than humans and they are part of our teachers....They're really really patient. They are from another planet, their original core, so they won't die out. They may die out on this planet, but they won't totally die out, and they won't die out anyway. It'll all stop before they all die out, but they are our teachers and they are observers, too. They're so incredibly advanced that what they have is a sense of humor. They're so far beyond us in intelligence, it's like no sweat. For them the message at the end of it all is very simple: have a lot of self-confidence, be who you are, love yourself, live in joy, take care of the earth, and have fun....

There's going to be lots closer bonds between animals and people. Definitely the whales aren't going to die out and definitely the animals aren't, and that's one reason that some of the destructions are coming by water. Because destructions by water don't hurt the water creatures....See and there really isn't global warming, that's really interesting; there really isn't, it's normal global changes. We are in a warming period and in many ways we caused it, but that warming period is just a natural cycle of the earth....I don't see everyone being a vegetarian, either. I don't see that happening, but I do see more respect for what you kill.[20]

Religions will also change, after a period of increasing repression, to a more earth-based, personalized type. There will still be the traditional religions but practiced differently. The Goddess religion will still be there, as individual and personalized as it is now, and with an important, quiet influence. The Goddess will never be a mainstream religion, and that's okay.

Eventually what we get to is very personalized religion, where there aren't very many great leaders. You'll have people building little community churches or worship places and some of the religions will last. There'll be a form of Christianity that lasts, and there'll be a form of Judaism that lasts, and there'll be a form of Zen — they will last. They will last in a form but they will all start to integrate more of a worship of the earth because the earth has almost destroyed them.[21]

The Goddess religion will be fine. It's never going to be mainstream. I think we're going to find an earth religion, sort of an earth kind of religion. People might call God "he" and that's okay, and the Earth Mother recognizes that. But the Goddess religion will be the permeating force that kind of keeps things moving. It still might be the minority religion but it will be okay, because it will be a religion that's more accepted....And it's okay with the Mother that it's a minority religion because simply those people who believe it and live it and create that essence of that belief system and relive that mythology — that is the energy that's very very important. It may not be important for those people to be mainstream and to spread their message to other people, it's simply important for those people to exist and to help affect the whole by the message they give.[22]

Tanith sees an appearance of Mary, the Earth Mother, as having worldwide importance.

I also think there's going to be what some people will call as Second Coming. It won't be Christ, it'll be a Mary experience and probably in South America, where we have another Mary Magdalene that shows up like we've had through the centuries. And part of her message will be, "I am angry. I am angry because you never listened. You listened to my son, you didn't listen to me, you twisted it all. I was the one, it was not my son. I told you to love and I let you borrow my son to see if you could do so and you couldn't. I told you to be joyous and you didn't listen." She'll be coming back with an angry message.

Diane: This is like appearances of Mary? Like at Lourdes and Fatima?

Tanith: Right, yeah. And probably in South America because there it'll be believed. It's the first place where that will be believed, so that's where it'll happen. The Mother is angry, and you'd better understand the Mother. Probably some Indian women and Black women and other women of color — it won't be white women — I mean it might be one, but it won't be all white women in these appearances. It'll be women of color.

Diane: The Earth Mother herself will appear as a woman of color?

Tanith: Yeah, because we have to deal with the Black Isis, the Black Kali, with the dark, secret, powerful side of the Mother. This will be the Crone aspect of the Mother and beyond that the primitive aspect of the Mother that says, I can also kill. And I will do so to save lives. I will do so to save the essence of what this is.[23]

When women stop listening to men and listening to men's power-over governments, they will take their power. Women's status in the New Age will change considerably from what it is today. Women's growth will be the growth of civilization on the planet, as it has always been from the beginning. In the women's New Age, for the first time since the patriarchal take-over, it is women who will be dominant and listened to. Women's ability to transmute pain into joy will transform the world.

It will finally be women who have come through enough cycles and turmoils and changes to be strong women, and to be secure women and to be who they truly are, and to transmute so much of the pain into joy and so much of the hate into love, and so much of the energy that's around into healing energy. These women transmute it through themselves, can actually be transmuters, so that energy comes in and goes back out in a different form, simply transmuted, and they are not harmed. They will transmute it for the earth so that healing energy is transmuted into the earth, transmuted for others around them. [24]

Women's networks will be highly important, comprised of energy transmissions from woman to woman that circle the world. One woman tells another, by letter or phone or telepathy, and the second tells a third. In this way, information and energy become universal and all are informed. If action is needed, what to do is transmitted through the networks. Whether the media has the story or the government releases the information is unimportant. Women will know through the networks. The most important connections may happen by "accident" and the women themselves may not even be aware of their part or of the network.

The women who are networking to network, that's not where the right connections are being made. That's okay to do, but the right connections are almost made by accident....The ones that happen, that just happen, will be the ones that need it and will weave the link that is to be established.

And they may never know it; they may never ever know it. They may never know that they're part of a psychic link that is connecting or the importance of it. It may not be necessary for them to know....

One of those women might be a little old lady in Craig, Colorado, who doesn't know anybody, but has two friends. One's in California and one's in New York, or one's in Montana, and she's part of that network somehow. She doesn't know it. She just knows that it's important to listen here and send it there. She has no idea why, and that's okay. It isn't important for her to know; it's important for her to be part of that connection, that network that creates that other energy. [25]

The re-emergence of women will mean the re-emergence of women's psychic knowing. How women create and live in a matriarchal world will be based on these psychic skills. The networks of women's energy that will transform the planet are transmitted by psychic energy. When the woman in Craig, Colorado, hears something and passes it on to her two friends, this is what she is doing.

Just knowing that this is the message and here's what we have to do. And letting those messages circle the earth so that as messages of healing and love and ecology and growth circle from woman to woman around the earth, they create an energy field around the earth on the psychic level, and that itself is healing for the earth and affects everybody else. Kind of like when people are under an electric wire something really does happen whether they know it or not. So if they're under a psychic field that's communicating this from woman to woman (and man, too) but all the psychics around the world, and they're all communicating that to each other, it creates that kind of energy field. [26]

The energy transmissions of women's matriarchal values will heal the earth and the people. It will be women's values of "harm none" that travel all around the world. A few men will be part of the networks, but very few; the energy is mostly that of women.

We might find a real upsurgence of almost like the Delphi prophetesses. Not gurus, more just that awareness that somebody who could be like that is there and sometimes we need to touch that. Those messages may not get spread in the newspapers, but they would get spread via that energy entering the world and then disseminating around the world. It'll be energy transmissions and we'll find that many women — and men, too but women are really the leaders of it — are going to be really incredibly psychic. Not psychic like great tarot readings, but psychic like talking from Pittsburgh to California and knowing, just knowing.[27]

Tanith warns of the ego danger to women who set themselves up as avatars, who separate themselves from the networks of sisters.

There'll be a danger for women of becoming avatars, an ego danger for women and we're going to see that intensifying, much like the dangers right now inherent in Craft for high priestesses where they get too egotistical. There's going to be a lot of people listening to women because they will be the ones transmitting this message and have transmuted that energy, and somehow their lives come out of the morass and become something beautiful and wonderful like a phoenix rising again. People will be attracted to them, and there will be a danger that'll be one of the other tests for these women. It's alright to have some ego about it because that's okay, and it's alright to even live by this, if you are able to support yourself by this, because that's okay. It's *not* okay if you do not become one that is connected with the others, with the other sisters. If you are not connected to the link of the sisters, you become only yourself and recreate that whole problem again. There will be a danger there for women.[28]

Along with the energy networks and the reclaiming of women's power will be the healing communities and schools that a few women will be part of. Most women will not choose communities of this sort, but some will, and these communities will have great influence in medicine and healing. The communities will also be Goddess-based, central sanctuaries of the religion of the Earth Mother. For the women who have been the most abused by patriarchy, these healing communities will be the most important. Alternative healing will become too prevalent, too much a force, for standard medicine to stop. It will exist side by side with technological medicine, and the women's networks will protect the healers from persecution.

Many of the network women, the connectors and communicators, will remain teachers without bearing children. They will make this a conscious or nonconscious choice, and will have their children in some other form.

It was a personal sacrifice for many of them. Many of them thought they never wanted kids but on some level it really was a personal sacrifice. Because there are many who need the training and they could not have the children, have the psychic sense, and be the teachers. They couldn't do all of that....

But they *will* have children. They will have children who knock on their door, and children who come, and children who are foster children, and they will be in the end fulfilled and have those children. They may get them when they are sixteen or seventeen or twenty; they may be fifty and get a twenty-year-old, but that bond will be there and it will be their child. And that training will be there, that teaching will be there.[29]

These are the women who will create the women's New Age, the Age of Women, and begin an earth culture of a thousand years of peace. These are the teachers and networkers, the leaders of a new planet. They are you and me, the women.

Perhaps the most interesting and exciting information from Tanith's reading is her material on the Dark Isis, and the pockets of energy being formed by the Mother under the earth. In the present, women are watching the disintegration of the current age. Much seems to be losing ground, seems to be lost forever, and various places on the earth are experiencing great desolation. There is an explanation for this in the emergence of the Crone aspect, the Dark Mother, and with understanding there is also hope.

There's like a pocket in the earth forming that's being filled with the energy of the Great Mother. She's pulling all of her energy to one great pocket. Two great pockets. One is in the ocean, under the oceans, and that's a safety pocket. That's a pocket of all the spirit essence of the whale, the spirit essence of the dolphin, the spirit essence of the hawk, and the eagle, and the elephant, and the whole endangered list. The spirit essences are being saved. It's like Mother Earth's ark, and it's under the ocean. It's a pocket, it's very safe, if all die she will create those again.

The other pocket is where she's putting all of her essence. Her essence has always been all around the world. She's pulling it in, she's pulling it into a pocket, and in that pocket will be her center of energy and her ability to make the changes, to really push the changes that need to happen on this planet. She has a place where she's saved in safety all that she might want to recreate and a place for herself. Because at this time she is becoming the Dark Isis and the Dark Kali, and the black side — which isn't evil but it's the powerful side. It's the powerful side of that woman/ Mother that says, I can kill to save. And I am also death at the end of all things, even though I am life at the beginning of all things. She is becoming that essence to move these changes, to move them, and so she is pulling herself in. Where she is leaving behind is where we're seeing great desolation, and that is one of the ways she's doing the changing. By leaving, pulling her spirit from a piece of land, from a place, and saying, you don't have me now and do what you must. What's happening in those areas is the desolation and the destruction.[30]

The earth changes are reflecting so deeply in individuals now, in women particularly, because women are so tied to the Earth Mother's energies. Along with the planet, we are facing and becoming the Crone, the dark aspect of the Goddess. With it comes wisdom and ability.

> So we feel that great shift in the Mother energy, and that the Mother energy is now becoming that dark energy. It's almost like you've been a two-year-old and your mother's really lost her temper now. That's part of what mothers have to do is once in awhile tell you that you've really gone out of line. And that's about what she's going to do. She hasn't lost her loving warm aspect, and this isn't an evil, nasty, cruel, punishing aspect. It's an aspect that says, I will not let you destroy everything. I will pull to safety what needs to come to safety, and let those of you who cannot make the changes and cannot understand live on the land that does not have my energy in it. And that land will not survive, because it doesn't have her energy in it.

> There's still that good side, there's still that loving side — and it's not good or bad — there's still that Maiden side and the Mother side and the mothering, nurturing side, and that will always be there through everything. That will still be there for us to tap. But part of what is being done is that the death side — the death aspect, the Crone aspect — is taking over a little bit and saying, uh uh. Anything that does go extinct she has that chance to reintroduce, and it'll be one of those miracles like we've already seen, where we thought something was extinct and then we found three or four somewhere, where they shouldn't have been. She's been doing it for awhile. She has that capability.[31]

This is never to say that the Goddess has rejected or forsaken her women in these difficult times of change. On the contrary, women creating the new ways and New Age are among those coming to safety. "The Mother has always been there for us," and we need to call on her, now as much as ever. Or perhaps more than ever before. It's a matter of women coming to understand the Crone, how to embrace her and to become her, and to pass through her to the Maiden on the other side of the curtain.

> It's a test. It's a test to see who can look at their issues. Who can look at what's really really happening and what they really do, and get beyond it and move forward into the next stage. Right now her concern has shifted away from helping us all through to saving the planet, so it's almost like we have been children for a very long time. The Mother's always been there for us, and now Mother is saying, Look I'm really busy and you need to go take care of this yourself. And for the first time we are not without her, but more on our selves. When you call on her she will listen, but you need to call. She's doing other things right now and her focus is somewhere else. She remembers who calls, and it's really good if you call. I mean, if you call it's not against you, it's an acknowledgement of, I know you're there for me still.

> But it's happening to people because she can't do people and animals and planet and everything all at the same time. It's gotten out of hand and she needs to focus. And because her aspect is more the Crone aspect, which is the wisdom aspect, we are all now heading very quickly from that nurturing aspect to that Crone aspect. We're not aging physically, but we're aging maturity-wise as we go through it.

Aging is a growth process and it's a painful process and so we're going through it quicker. We're going through this period of her bringing out her Crone energy and the earth having its Crone energy. After that we'll go back to the Maiden. But we have to go through the Crone, and the Crone can be scary.

For some people it's like the curtain: a few days ago I saw what was on the other side of the dark curtain, and all it was was light. It's the dark curtain that is the Crone and it's the light that is the Maiden. And it is simply the movement through the Crone that lets us grow and mature and be who we need to be, and be reborn as the Maiden.[32]

On a lighter note and after the earth changes are completed, Tanith talks about women with wings.

Here's my other side kicking in, my other side is saying it just got this message that's so bizarre, *I'm* not going to say it! Somebody might read this book, and they're going to think I'm crazy! Women are going to have the option of developing wings again....

I think I should write about that to AmyLee because she said something about the Indians know that women had wings. Women will be able to very very slowly, very slowly, over time find their wings again, develop their wings.

Diane: Is this teleportation, or is this physical wings?

Tanith: I'm getting physical wings....I'm getting wings. Your body could have something attached to it that's not seen in this dimension, and those wings could be in that other dimension. They're really attached to you, but no one gets to see them, kind of like auric wings or etheric wings — you know, that other-part-of-you wings. Women will have the chance to get those back, if we want them....

That'll be a gift, that'll be a gift for those women.[33]

Tanith plans to be the first on the signup sheet for wings — and I will be next in line! Laurel Steinhice, in her channeling, describes a planet where women have wings, also. For more on the role of men in the earth changes and the New Age, information on contact with other planets, and on the coming future for earth regarding dis-eases and the attitude toward death, read Tanith's full tape transcript in the Appendix.

Laurel Steinhice speaks in her channeling of a New Age that is an increase in dimensional frequency for the earth and its people. She calls this change the fourth dimension — what Sheila Petersen-Lowary calls the fifth dimension, and what in Hopi cosmology would be the fifth world. She describes this fourth dimensional earth as a physical place with more than physical abilities. Tanith's discussion of women with wings is a good lead-in to understanding this information, channeled through Laurel by Edgar Cayce.

It will be a new kind of physical. It will be much like the spirit plane is now. There is nothing available to you on earth planet that is unavailable to us (discarnate Be-ings). If we want bodies, we manifest them, complete with all senses: taste, touch, and so forth, sight and hearing. If we want homes, they appear. It is creation of personal and interpersonal reality. There is nothing you have that we cannot have. You want sex, you want drugs — if you really want it, you can get it on the fourth dimension. Let us be honest about this.

Diane: Is there a difference in suffering?

Edgar: The suffering is here, yes. There are fourth dimensional planes of suffering also, but this earth will not be one of them. And no one will go there except by their own choice. But in addition to everything available to you, we also have certain nonlimitations. Communication, telepathic communication is universal. Telekinetic travel and transport, also. We have only to think where we wish to be and we are there. This you are already learning to do, you and many. [34]

Edgar defines the difference between the current third dimension and the coming fourth dimensional reality:

What is third dimensional reality? It is consensus reality. You have heard of "We create our own reality"? Yes, personal reality is interconnected with that of others. The consensus reality for this planet has been redefined. It has already been chosen and is in the process of being manifested. It is called fourth domensional vibrational frequency. The new earth will be physical, and the bodies on it will be physical. New kind of physical earth, new kind of physical bodies. You are in the process of adjusting the body as well as the planet for this new physicality, this new consensus reality, which we call fourth dimensional frequency. [35]

The psychic abilities of teleportation, astral travel, telepathy, and contact with entities or discarnate spirit guides are fourth dimensional abilities. Many women participate in this reality today. While in the past it has been necessary to die and to leave the earthplane completely to enter a fourth dimensional reality, beginning today and in the women's New Age things are different. Instead of leaving the body behind to make this dimensional shift, the body is being prepared to come along. In time the whole planet will enter this fourth dimensional reality.

At this time, it is not necessary to die to go from third to fourth dimension. You are already doing this, as many are, and you are adjusting your body so that the body can move directly into fourth dimension without experiencing a death crossing transition, without being left behind.

Diane: The body will go too?

Edgar: Yes. And will still be able to function in third dimension as well as fourth. This is what you are doing now. You are swinging back and forth. It is not from low third to high fourth, they are close together. And you are moving from high third to low fourth. There is much interchange between, and lightworkers, many people, everyone who chooses, need not die, but can take the body directly into fourth dimensional frequency. This is why you have fitness craze; they are preparing the body. This is why you are releasing old blockages; you are clearing the way. And in this high third/low fourth, the earth has come into alignment with the lower astral planets. And this is why disturbances, spirit attachments, and all sort of old garbage is manifesting, for the purpose of being healed.[36]

The earth as a whole is more than halfway into this new vibrational frequency already. By the time of the axis shift most of the lightworkers will be comfortably living in that fourth dimension and the earth will make her final transition into that energy. Tanith put it in another way when she said that the earth is in the phase of the Crone, and that on the other side and through the dark curtain, she becomes the Maiden again. Women, of course, are evolving with the Goddess earth. Says Edgar through Laurel Steinhice:

We're all in this together. We ride the storms, we adjust the energies, we commune and communicate with each other. Very many are engaged in this.[37]

This is another way of describing Tanith's psychic networks and the women who are the connectors in them.

The addition of fourth dimensional skills to women's psychic senses has some interesting ramifications for today:

Let us talk about some practical realities. Think how useful it would be when the lava flow is coming down your street if you can levitate. You raise yourself above it and it flows by underneath you, yes? But you need not wait for the lava flow to come down your street to find levitation useful. Let us assume there is some very physical, mundane, earthplane harm and you can levitate. But you need not raise yourself all the way above him so he runs by underneath — you get this far off the ground and he becomes alarmed and will run away.

Many of those tricks — mystical, magical, theatrical tricks — are practiced for the future. Levitation, dematerialization and rematerialization: we're talking about fully functional ascension and interaction between dimensions....

The body goes, too. It is not only the spirit, it is the body also. You are manifesting fourth dimensional vibrational frequency by your spirituality.[38]

Levitation, dematerialization and rematerialization, astral travel with the body brought along, psychic networks, telepathy, teleportation, and women with wings. The women's New Age will be very interesting indeed. After the time of the Dark Isis, the Crone aspect, the light on the other side returns the Maiden. The fourth dimension or women's New Age will be a place of women's "harm none" values, a different world in an Age of Women — and it sounds like a lot of fun besides. Women's thought power and our participation in a matriarchal consensus reality are bringing it into Be-ing.

Notes

1. Starhawk, *Truth or Dare*, p. 26.
2. Mary Summer Rain, *Phoenix Rising*, p. 134.
3. *Ibid.*
4. *Ibid.*, p. 154. I have changed "man" to "people" here.
5. Marion Webb-Former, *On the Future*, unpublished.
6. *Ibid.*, p. 2.
7. Tanith, *Reading, May 7, 1990*, unpublished.
8. *Ibid.*, p. 7.
9. Tanith, *Reading, May 19, 1990*, unpublished.
10. Tanith, *Reading, June 21, 1990*, see Appendix, p. 212.
11. *Ibid.*, pp. 210–211.
12. *Ibid.*, pp. 211–212.
13. *Ibid.*, pp. 214–215.
14. *Ibid.*, p. 214.
15. *Ibid.*, p. 214.
16. *Ibid.*, p. 220.
17. *Ibid.*, p. 212.
18. *Ibid.*, p. 225.
19. *Ibid.*, p. 213.
20. *Ibid.*, p. 218.
21. *Ibid.*, p. 216.
22. *Ibid.*, p. 217.
23. *Ibid.*, pp. 216–217.
24. *Ibid.*, p. 217.
25. *Ibid.*, p. 223.
26. *Ibid.*, p. 217.
27. *Ibid.*
28. *Ibid.*, p. 223.
29. *Ibid.*, pp. 225–226.
30. *Ibid.*, p. 220.
31. *Ibid.*, p. 221.
32. *Ibid.*, pp. 222–223.
33. *Ibid.*, p. 227.
34. Laurel Steinhice, *Earth Changes Channeling*, June 3, 1990, see Appendix, p. 197.
35. *Ibid.*
36. *Ibid.*, p. 206.
37. *Ibid.*
38. *Ibid.*, pp. 206–207.

Samplings of Realities

Women's Utopias

Yet there are those who wonder. There are those who have gentle stirrings. And there are those who have stepped upon the beautiful threshold of awareness — all on the verge of perceiving that which is there to see. To these ones, I say, open your exquisite senses. Look with fine clarity into that which is beyond and beneath, within and without. In these coming critical times, listen to and heed the directives of your spirits that retain the high wisdom you are just now perceiving.

Mary Summer Rain[1]

Women are creating the New Age, the Age of Aquarius, the Age of Women. By our consensus reality we are creating it right now. Every time we step into the astral to visit a guide, foremother or living friend, every time we do distance healing, a tarot reading or divination, and every time we "just know" and honor that — we are entering the New Age and creating it. Every time healing is done for oneself or another, healing for the earth in her changes takes place. Every time we expand personal reality, release an inner block or resolve an old issue, we are healing ourselves and the earth, and entering the Age of Women. We are already more than halfway there.

Thought power is the vehicle for this creation. There are as many possible realities as there are women whose fertile minds and imaginations (and divinations) create them. It is already clear that we have the power to manifest what our visionings design. Manifesting positive possibilities, rather than negative horror stories, is vitally important. The world is what we think it to be; it will become what we think. We can think it into reality.

In the time of the earth changes, women's healing, lightwork and positive energy have the power to save many lives and prevent much devastation. Whatever occurs, it would have been worse had there not been women working toward healing and to ease the birth pangs of the Mother. As we pass through the time of the Crone, the Maiden is on the other side, a newborn phoenix ready to meet the light beyond the dark curtain. These mixed metaphors are symbols of the process of the changes, of the planet's transformation, and of women's ability to transmute pain to healing and joy as the process unfolds.

After the earth has changed and the patriarchy has disbanded and dissolved, people will look to women to guide the way to a new order. A women's order based on matriarchy, Goddess/women's values, respect for all life, and "harming none" will right the current wrongs and herald a thousand years (or more, we hope) of peace and abundance. The people of the earth have had enough of patriarchy's power-over, and women and men alike will look to the feminine for other ways of living and governing. We are the women of that New Age, the aware women who understand the earth changes and women's part in them. It is up to us to create that new world as a positive place, a place where we'd want to raise children.

With this in mind, what kind of world will we create? If anything in the information of this book is disturbing or hard to accept, work to find alternatives and change it. Think the alternatives into being by visualizing, meditating, dreaming about a perfect, "harm none" world. While thinking the alternatives, do activism on the earthplane as well — in the voting booth or on the protest lines, by writing editorials or teaching others awareness, by recycling and living the Earthway. Create a vision of the women's New Age and talk about it with others, depicting an age of peace and well-being for all. Create a vision, tell it as a story, write it as a book, paint it as a picture, sing it as a song. Create a vision, and think that vision into Be-ing.

Women have always been the visionaries for the realities and worlds of the future, and women's literature is filled with examples of possible new worlds. Some of the earliest of feminist fiction is utopian fiction, as in Charlotte Perkins Gilman's *Herland,* a women's utopia of 1915. Women have always looked at the world with an eye toward improving, to righting the wrongs and creating a new dream, a new way of living and society. Where male visions are filled with wars and conquests, women's are filled with peace, growth, fertility, cooperation and wellness — what we want to manifest in the New Age.

Modern women's fiction, speculative fiction or science fiction continues to create these new worlds as places to try out the future. For a few hundred pages, anyone can live in a women's utopia, a women's New Age and experience it. Possible realities are explored and temporarily manifested, to be rejected if unworkable or uncomfortable, and to manifest by thought power if we choose. By examining some of the realities, some of women's created new worlds, a variety of choices is presented for what to manifest.

The women's fiction of this chapter was published between 1976 and 1990, and is therefore quite recent, but also representative of the current women's movement. The writers may or may not be aware of earth change theory, though some of them seem to be. All of them are aware, however, of the intolerableness of the present order and the need for changing it in "harm none" ways. Some of the women present women-only and lesbian separatist visions, while others are heterosexual. All but one of the novels presumably takes place on earth, with one happening on the earth colony of Venus. Some give no timeframe and could be happening at any time including now, while others are in the definite future. All the novels are feminist and affirming of all women and all races, as well as containing an awareness of the earth. Some of the novels are Goddess-based, while most are not. All are written by women and feature strong female protagonists that are role models. None are dystopias, focused on the disintegrating world, but all are positive examples of possible realities in earth's future.

Sunlight's *Womonseed* (Tough Dove Books, 1986) is perhaps the simplest and clearest of these fictional visions of a world as women could create it. The book takes place in 1999 at Summer Solstice, with the women of the commune gathered around a fire to tell stories. The older of these women left the modern world, one by one, and each in her way came to Womonseed to start a new life living with women and the earth. The children of the circle, some born by parthenogenesis (conception without the male), hear the stories of their mothers' world, a world foreign to them, and offer stories of their own lives. These New Age children are innocent, unhardened, and spiritually aware.

The commune was started by one woman, who upon leaving her husband takes a woman lover and uses divorce settlement money to buy the land. The women begin to come, and when a forest fire destroys the house on the property, they learn to live in the Earthway, rejecting electricity and motor vehicles for a more simple life. An aura of great love and caring permeates the colony, a place where women and girls of all races, ages and abilities learn to care for each other and live well. There is the feeling of an oasis, a place of peace and safety away from the craziness of a dangerous male world. Many of the women come traumatized from that world to Womonseed, and find healing and fulfillment.

Their journeys led them through forests of dense shade, over mountains deep in snow, across deserts shimmering in the sun. Somtimes the path was hidden in weeds, obscured by tangled underbrush or covered with wind-blown dust. Then they held each other's hands, listened to their inner guides, and found their way again. When the women were hungry, the plants along the road offered themselves as food. When they were tired, the earth held them gently in her strength. When they felt sad or afraid, the women took each other into their hearts and arms until their courage was restored....

And so for many cycles of the sun the women travelled, searching. Some of them came to a beautiful valley — long and sweet, like the long, sweet valley between a woman's thighs. They touched the earth and knew that this was home. They celebrated, singing and dancing day and night, while the hills echoed with their joy. This was a land full of promise, like their hearts. They felt free, for at last they had escaped from their captors. They felt hope because they saw the moon and the sun and rainbows shining with beauty, lighting their new world. They felt safe as the valley embraced them. The celebrations went on and on until the first winter rain began to fall.[2]

The women build shelters from the winter rain — this is a gentle California climate — and learn to eat from the forest and plant food crops. It is a vegetarian commune, where the animals of the forest are respected and safe. The women create the culture with their own hands, separate from the outside world. This is a lesbian, Goddess culture, a place of healing women from patriarchy's wounds, and a place of renewal and hope for the next generation. Storytelling is a form of entertainment, as well as songs, meditations and rituals.

> Life is simple there because it flows simply out of being. The women and children get up with the rising sun, their dreams mixing slowly with the light. They plant the fields and gather from the woods. They bathe in the creek and dry on its banks. They build shelters from the rain. They make pots from the red clay soil, weave baskets from reeds that grow by the river's edge. They make love in the soft, clean air. Together, they eat simply from the bounty of the earth. Then they gather by the fire to sing their songs and tell their stories.[3]

Each woman at Womonseed has her story of where she came from and what her life had been like. The stories are typical of the sufferings of women in a repressive, disintegrating patriarchy. There are the tales of bad marriages, abused children and wives, dangerous jobs that are the only jobs available, repressive office situations. There are stories of courage, how each woman left her situation and found her way to Womonseed, often in great danger and in need of healing. Kahuna, called to the land from Hawaii, has a vision of the future, a vision of the coming earth changes:

> There it was, what we have seen — the wars, floods and droughts, famine, plague, volcanoes, earthquakes. It showed me what was coming down, and I intended to survive it and knew we could — some of us. I knew a lot by then. I knew that we have power and I wanted to share that knowledge with other women to use to keep our lives and to make them good. For the first time, for many of us. I saw a center, a community, an oasis where women could go — not to escape, but to start a new kind of world. A world based on the best in ourselves, based on knowing our power and creating what we want from that source. I started visualizing that image consciously and often, and the vision grew. I saw a gentle, fertile land opening between high hills, with plenty of water — a creek, a pond, clear, sweet water flowing out of springs. I saw a circle of women and children singing songs of creation. I saw them workplaying in the fields and sleeping out in the night. I saw the love that wove them, weaves us together.[4]

Before coming to Womonseed, Sassafras also had a vision. The women thought their visions into Be-ing, the creators of this new and gentle world. Sassafras heard a voice, and Moonstone heard it with her. They came to Womonseed and made the vision real:

> You will found a center of love which will be one of the bridges between this age and that which is to follow. It is an age of love and understanding among all beings.

> Many will be teachers of this age for many will be needed. Cataclysmic events will shake the earth. They are doing so now. Do not be afraid of them because these are forces clearing the way for the new age. Part of the earth is destroyed, many people "dying." It may appear sad. There will be pain before acceptance. The pain is the resistance to the new. It is always so. Nothing, no one is lost. It is a transformation.[5]

Each of these women finds her way to Womonseed, to help build the women's New Age and tell her story at the Summer Solstice fire.

The New Age children of Womonseed are the pride of a new world, too. They are cooperative and loving, open and independent, strong and self-assured. They are aware of the Goddess within them, and unblighted by today's women's blocks and wounds. These are children raised on the land and raised with love. Womonseed's daughters are telepathic and spiritually aware. They have psychic abilities and senses that their mothers watch and wonder at.

Oya is blind, using her psychic sense to guide her. She leads her friends Thistle, Sierra and Manzanita to water in the woods with an awareness that the others lack.

> "Touch it, Oya. There's no water in it."

> "It's coming. Can't you see it?"

> Sierra took her hand and mine and so did Manzanita. We circled around the stones, focusing our energy there. With closed eyes, I made myself see a spring bubbling out of the sand. Clear, fresh water that looked and tasted like the snow-brook on Mount Shasta. Soon I could hear it too — water sounds tumbled the pebbles as it flowed. I splashed it on my face and soaked my feet. It became as real as any water I'd ever known. Then I felt small, wet fingers touch my mouth, and I opened my eyes. Oya was dipping her cupped hands into the spring, feeding water to us. I kissed her and thanked her and the Goddess for showing me how to see. I felt we should have some kind of ritual before we quenched our thirst, but Oya was giggling and calling the horses over. I realized the ritual had already taken place. I laid down on my belly and drank deeply.[6]

The women's visions of a better world create a place of peace and love in *Womonseed,* changing the earth change era into the Age of Women. The old, the new and the future are presented in this gentle, thought-provoking and beautifully written version of women's future.

One of women's earlier visions, and a book much loved by many women, is Sally Miller Gearhart's *The Wanderground: Stories of the Hill Women* (Persephone Press, 1978). In this world that has some similarities to *Womonseed,* there is more contact with the dying old order, less separation and utopia, but *The Wanderground* is a utopia established and building. The women of this settlement are highly psychically developed, and base their lives upon it. The old world is a continuing clear and present danger, and the hill women use their psychic senses daily to save what they can from it. They are a bridge between the old and the new, as are the women of *Womonseed,* but there is more of a feel for action and process here. The old is not dead but is dying, and the new is emerging and being born.

Unlike *Womonseed,* men are a factor in *The Wanderground,* and they are problematic figures. There are the men of the City, caricatured patriarchal stereotypes, as objectified as they themselves objectify women. The City is the patriarchy at its most extreme, male clubs and places of danger for women. The hill women go there disguised as men to gather information and to help women when they can. One works in a hospital rescuing female infants with psychic ability that are to be euthanized. The City is a place devoid of love, and the women there are damaged by the harshest abuse:

> Ijeme looked at her unbelievingly. She was a thing out of history to the hill woman: a thickly painted face, lacquer stiffened hair, her body encased in a low-cut tight-fitting dress that terminated at mid-thigh; on her legs the thinnest of stockings, and the shoes — were they shoes? — Ijeme could not believe they fit the same part of the anatomy that her own boots covered. How could she walk in those spindly things? And with the flimsy straps that fastened them to her ankles and feet? The dangles that hung from the woman's ears jangled in tune with her bracelets. She clutched a cloth-covered purse to her side....

> This was the city edition, the man's edition, the only edition acceptable to men, streamlined to his exact specifications, her body guaranteed to be limited, dependent, and constantly available. Ijeme shuddered, then repeated to herself the words of an early lesson, "What we are not, we each could be, and every woman is myself." [7]

When this woman dies, no one cares but Ijeme.

The other type of men in *The Wanderground* are the gentles. These are men who have forsworn the privileges of City males. They want to work with the hill women, and sometimes do, but the women are suspicious of them. Many of them are gay, and many of them are dying — with an unnamed dis-ease that predates but prophesies AIDS. The hill women, damaged by the negative culture from which they escaped, cannot help them to die or to live.

> Gentles. Men who knew that the outlaw women were the only hope for the earth's survival. Men who, knowing that maleness touched women only with the accumulated hatred of centuries, touched no women at all. Ever. Once, she remembered, some gentles had come to the Wanderground, stricken and dying. Unwilling to return to the City where they might have been revived, they came to the hill women. They came for help in their dying. They cried for the ministrations of the women. [8]

The gentles are allies, as damaged by the City as the women have been, but unable to be accepted by the women, either. The Wanderground is a harsher reality than is Womonseed; it is still embattled for survival.

She tried to recall the lessons from the remember rooms: the stories, the mind pictures, the pain of some not-so-ancient days when the men owned all things, even the forests and hills. "It is too simple," she recited dutifully to herself, "to condemn them all or to praise all of us. But for the sake of earth and all she holds, that simplicity must be our creed."[9]

The life of the hill women is both simple and complex. Their abilities at telepathy, far-seeing and sensing each other makes them close-knit in cooperation and living together, but highly independent and self-identified, loving but emotionally distant. The women talk with each other telepathically from almost any distance, and also talk with their animal companions, which they regard as equals. They live in oneness with the elements and the earth, and the land they live in is wildly beautiful.

Rabbits. Or some small animals just below the cold ground. Above, strangely quiescent starlings. Or sparrows? Around her, fallen branches, deep moss, damp grass, red-brown mud, dormant brambles, layer on layer of thicket, the sun passing behind the far rise, the river moving slowly by and swirling faster beneath the giant tree, the far-off promise of a midnight forest. Quietly she swung her stretch further to full circle at a distance beyond the rise. Less intense sounds and smells now, but more of them. By swift montage she listened to and felt one at a time, every thing, every oxygen-breathing thing, every other-breathing thing, every non-breathing thing. They felt her attention and told her all was well.[10]

The Wanderground women, or some of them, have also learned to ride the wind, their psychic sense carrying them aloft and over distances. Remember Tanith's women with wings, and Laurel Steinhice's description of women's fourth dimension psychic skills.

How long over the quiet road she sped and braked, drifted and dived, Evona did not know. As she rehearsed paradigms of movements and their variations, she was aware that she was waiting for something. She could not name it until it happened and when it happened, she named it aloud: "Footloose!" The change had come — the shift from being earthbound to being windborne.[11]

These are women whose abilities have exceeded our current third dimensional earthplane. They are psychically mature.

The earth herself has turned the order of power from the men to the women, but the change-over is still incomplete. On one side is the City, filled with abuse, technology and fear, and on the other is the community of women, struggling to hold onto their freedom and culture. There have been persecutions and "cunt hunts" that left many women dead. Children are taught in the remember rooms about the Revolt of the Mother, the day the earth said no, the day there was one rape too many:

There was no storm, no earthquake, no tidal wave or volcanic eruption, no specific moment to mark its happening. It only became apparent that it had happened, and that it had happened everywhere.[12]

And "it" continued to happen:

Machines outside the City...were working no better than usual. Breakdowns were still consistent — planes faltered after less than an hour's flight, trains and autos ground to a stop after short bursts of speed, sails and oars were still the only means of progress over water. Natural-grown food was still a luxury, the chemical substitutes still the standard. Communication with any other surviving city was limited to runners. Horses and mules and other beasts-of-burden still refused male riders or drivers.[13]

Males became impotent outside the City, and as their guns refused to function and their dogs refused to track down women, the hunts had stopped. The women learned to reproduce by cell fusion, bearing only female offspring, to develop their psychic senses, and to live with the earth in a new way. The survivors of the City became the new women, the new world, the women's New Age. *The Wanderground* transmutes pain into joy, and the changing earth into a newer way of Be-ing. It is a separatist vision and sometimes harsh, but always beautiful.

Suzy McKee Charnas' novel *Motherlines* (Berkeley Books, 1978) presents another utopia of free women. These are women who have escaped enslavement in the deteriorating patriarchal cities and fled to the desert where they have cut all contact with men and the women remaining behind. They are survivors of the Wasting, the end of the patriarchal era. One group, the Riding Women, base their desert culture on their horse herds and lead a nomadic lifestyle. They are a closed Motherline society, sending newer escaping women to the camps of the other group, the Free Fems, a collection of escaped slaves. There are major physical and psychological differences in the two groups of women.

Alldera, a slave, escapes the Holdfast to try her luck in the wilderness; the women's communes there are no more than rumored. Becoming pregnant via rape at city's edge, she is found by a Riding Woman who brings her to the Riding Woman's camp. Because she is pregnant with a daughter that could one day be founder of a new Motherline, Alldera is taken in by the Riding Women and nursed through her pregnancy. Her child, genetically readied, is made part of the Riding Women's society, given a family of sharemothers, and raised as one of the camp. Alldera feels alien, unable to adapt to a freedom she has never before known.

The Riding Women are wild and free, highly independent and with a developed society. The camps of the Free Fems are different, not a culture, but a group of women who though free still retain slave psychology and a hierarchy based on their former status under male masters. They are free in body but not in spirit, and unable to reproduce. The Riding Women bear daughters, partheno-genetic copies of themselves and descendants of their Motherlines:

> "Their beginnings and ours differ," Nenisi said. "Around the onset of the Wasting that ruined the world of the Ancients, there was made a place called the lab, where the government men tried to find new weapons for their wars. We don't know just what they were looking for, but we think it was mind powers, the kind that later got called 'witchery.' The lab men — and lab women, who had learned to think like men — used females in their work, maybe because more of them had traces of the powers, maybe because it was easier to get them with so many men tied up in war....
>
> "The lab men didn't want to have to work with all the traits of both a male and a female parent, so they fixed the women to make seed with a double set of traits. That way their offspring were daughters just like their mothers, and fertile — if they didn't die right away of bad traits in double doses....
>
> "The daughters got together and figured out how to use the men's information machines. They found out all about the Wasting, the wars and famines and plagues going on outside, and how the lab could be made self-sustaining if things outside collapsed completely. They laid plans of their own.
>
> "They got the information machines to give a false alarm warning of an attack in the offing and ordering the lab men to rush off to the Refuges and save themselves....The first daughters sealed themselves up safely in the lab and using the information machines began to plan for after the Wasting. They took the lab animals and tried to breed them to be ready to live outside when the world was clean again....
>
> "(The women) perfected the changes the labs had bred into them so that no men were needed. Our seed, when ripe, will start growing without merging with male seed because it already has its full load of traits from the mother. The lab men used a certain fluid to start the growth. So do we."[14]

The fluid is horse semen, and the women use their horses as the basis of their society, a society of free and equal sisters. A child enters the childpack at weaning to run free and untamed until her first menstruation. At that time, the other daughters chase her from the pack, and the child's birth and sharemothers ritually bring her into the camp to train as a member of the community. Having grown up wild, she becomes a very free adult. On maturity she is helped to trigger a pregnancy and bear a child genetically identical to herself. The Motherline continues and the Riding Women's culture survives the generations. The women share a deep bond with each other and their horses, living from their herds' products and the products of the earth.

While the Riding Women have developed a sister-centered culture, the Free Fems have not. Emulating the only rules they know, of power-over and competition, their lifestyle is much less positive than that of the Riding Woman, whom they despise. Their dream is to return to the Holdfast and free the other slaves, and to bring male children from the city so they can reproduce. The Riding Women prevent them from approaching the city, fearing their failure would endanger the lives of both groups. The Riding Women believe that the men have almost died out in the cities, and that they themselves are the beginning of a better world.

Unable to adapt to the Riding Women, Alldera goes to the camp of the Free Fems, where she is not well-accepted. Taking friends with her to live semi-solitary on the plains, she teaches the Fems to ride, learns to tame her own horses, and becomes a bridge between the two groups of women. When her daughter leaves the childpack and Alldera returns to the Riding Women, she brings Free Fems with her to learn independent ways in the women's camp. Now a member of two cultures, Alldera finds her place in the society at last, and the Free Fems learn emotional maturity.

The Fems worship a Goddess, whom they call Moonwoman, while the Riding Women believe in a pattern of existence that they have no name for. The Riding Women's bond with their horses is their bond with that pattern and the earth:

> We celebrate the pattern of movement and growth itself and our place in it, which is to affirm the pattern and renew it and preserve it. The horses help us. They are part of the pattern and remind us of our place in it.... [15]

Alldera's daughter is the child of both worlds, a symbol of the hope of a new world, a new culture and pattern, a place for women after the Wasting when wars, plagues and changes are over. The Riding Women are reminiscent of the survivor groups described in the future life progressions, living without technology in a frontier lifestyle. The Free Fems are women unable to adapt and move forward in a new world, only looking back at what they've lost. Alldera is the link between the old and the new, and her daughter is a child of the women's New Age.

Where the preceding three books are lesbian-separatist models of a technology-free New Age, the next three are heterosexual but openminded futures with more of a focus on modern society and technology. Sheri Tepper's *The Gate to Women's Country* (Bantam Books, 1989) provides a thought-provoking concept of how to create a New Age world without war. Her vision may represent one way, but is it the best way, for women's thought power to create a new society? In Sheri Tepper's culture, men and women live almost entirely separately with the exception of a few men who act as servants within the city walls. Most men live outside the gates, in an army garrison with a warrior's mindset, and a male life goal of conquest over other garrisons. Women live in this city, in Women's Country, a peaceful community, highly developed for medicine, teaching and the arts.

When a woman wants sex or to bear a child, she goes to an assignation house at one of the open-gate festivals, where she meets the warrior of her choice. If her child is male, he is sent at five years old outside the gates to be raised by his "father." At the age of fifteen, the boy must make a choice, to remain outside as a soldier or to re-enter the gates as a servant to the women — and be highly derided for it by the men. The men who choose the city and live in peace there are given education denied to the soldiers, and a far gentler, saner life. Unknown to even most of the women, the men who return are the fathers by implantation of all their children. The plan is to breed a group of men who deny war, who re-enter civilization willingly and in peace for a better New Age.

This story unfolds only slowly in the book, and is not fully revealed until the end. Morgot describes the earth change, brought about by devastating war, and women's determination that it will never happen again:

> Three hundred years ago almost everyone in the world had died in a great devastation brought about by men. It was men who made the weapons and men who were the diplomats and men who made the speeches about national pride and defense. And in the end it was men who did whatever they had to do, pushed the buttons or pulled the string to set the terrible things off. And we died,...almost all of us. Women. Children.

> Only a few were left. Some of them were women, and among them was a woman who called herself Martha Evesdaughter. Martha taught that the destruction had come about because of men's willingness — even eagerness — to fight, and she determined that this eagerness to fight must be bred out of our race, even though it might take a thousand years. She and the other women banded together and started a town, with a garrison outside. They had very few men with them, and none could be spared, so some of the women put on men's clothes and occupied the garrison outside the town....And when the boy children were five, they were given into the care of that garrison....

> When enough years went by, it was no longer necessary for the women to play the part, and it was left to the men. Except for those few who chose to return to the city and live with the women. Some men have always preferred that....

> In the first hundred years, the garrison twice tried to take over the city. But the women had not forgotten their years as warriors....They fought back. Also they greatly outnumbered the men. It is part of our governance to see that they always greatly outnumber the men....

> In the two hundred fifty years after that, warriors have tried to take over this city, or other cities, time after time. None of the rebellions have succeeded. [16]

Morgot's daughter is Stavia, the protagonist of the novel. A teenager at the beginning of the book, she reads about preconvulsion societies with her friend Beneda:

> "Reindeer," she said, half strangled by her own teary laughter.

> "What do you mean, reindeer?"

> "Just...we don't have them anymore."

> Beneda's mouth dropped open. "Stavvy, honestly. There's lots of things we don't have anymore. We don't have...clothes-drying machines and mechanical transpor-

tation and furnaces that heat your whole house, and cotton and silk and...and cows and horses and...and all kinds of other animals and birds and — oh, lots of things."

"I miss them."

"You've never *had* them!"[17]

The women's towns are not the only communities. On a trip south of Women's Country exploring the badlands, Stavia, now grown, is captured by fundamentalist members of a survivors community. The settlement is looking for women; the group is dying out because of overzealous euthanizing of female infants. These Bible patriarchs carry to conclusion all the logic of the crazy modern right-wing. Susannah houses Stavia. She is twenty-nine and worn out from pregnancy after pregnancy; contraceptives are forbidden, and the primitiveness appalls Stavia who tries to help.

> Susannah shook her head. "They might catch you again. Besides, there's no need. It's comin' to an end, can't you see? More'n more babies born dead or put out to die because there's somethin' wrong with 'em. It's all comin to an end, and I'm glad. It's just...you know, you get to love your girl children..."[18]

Stavia is rescued from Holyland by the male servitors of her mother's house; their psychic abilities "hear" her distress and they go to find her. Susannah dies by suicide, and her daughters are taken to Women's Country for schooling and freedom.

Perhaps the great flaw in Women's Country is its need for secrecy about the true roles of the men. Most of the women are kept ignorant of it to prevent the reality from reaching the warriors. The warriors are there primarily to die. Their numbers are kept down, and when they become too unruly a war is declared that decimates their numbers. Their other purpose is to be a testing place for the male children, to see which boys will return through the gates to accept peace. Each generation, more and more boys return, and the women are pleased; the warriors dub these boys cowards.

Before the open-gate carnivals, the women are implanted with contraceptives that are removed after the festivals. The women are told that these implants prevent disease. When they are removed, some women are inseminated with sperm from chosen men, servitors in the city. Genetic selection is the purpose; the women do not bear warriors' children, but children by men who have chosen to live with the women. The unknowing women, however, are both encouraged to go to the assignation houses and to remain aloof from the warriors. Younger women, not knowing the reasons, develop crushes that affect their lives and the city adversely, and this is worked out through Stavia in the novel. The women never think to focus their affections where they might be more rewarding, with the men who live in their own house as servants. Or with each other.

A traveler discovers the secret, as it is revealed near the end of the book:

> "There were other clues....Firstly, everyone said that more men came back through the gates in each succeeding generation. That argued for something, didn't it? Selection, perhaps?....

> "And then there's the matter of the servitors. Some of them, of course, are like Sylvia's Minsning, fluttery little fellows who are simply happier in Women's Country as cooks or tailors or what have you. For the most part, however, the servitors are more like Joshua or Corrig, highly competent, calm, judicious men, and they are highly respected, particularly by the most competent women. It argues, that both their status and their skills exceed what is generally supposed." [19]

Much grief could be saved many women by this information.

The Gate to Women's Country is another possible reality of the many available to women in creating a New Age. Whether this alternative is the only way to end war is unknown, but women have the choice of this or any other creation in their thought power. Hopes are that the great devastation of this book, or the Wasting of *Motherlines,* and/or the societal degeneration of *The Wanderground* and *Womonseed* do not have to happen, and that world peace can come in easier ways.

Surprisingly the oldest, and also perhaps the most modern, of the women's utopias and new worlds discussed here, is Marge Piercy's *Woman on the Edge of Time* (Fawcett Books, 1976). Its far-reaching vision makes it seem newer writing than it is. The novel connects the world of today with that of the future, the year 2137, and offers the message that tomorrow is being determined by today's actions and women. The future can be a utopian women's New Age, or a nightmare of the harshest repression.

Connie is a Chicana woman of New York City in 1976. She is wrongfully imprisoned, for no other crime than being poor and nonwhite, in a mental hospital. What makes her unbearable world livable is the series of visits she makes to the future, to the Mouth of Mattapoisett utopian colony, aided by Luciente, a woman of that time. The contact with Connie is made psychically by Luciente in an experiment of her future. Its purpose is to influence the past so that the peaceful, free, life-affirming community can later exist. By making women of 1976 aware that there are other ways to live, by planting the seeds of that earth-based and "harm none" way of life, Luciente hopes to make Mouth of Mattapoisett a future reality.

While all of the books so far, except for *The Gate to Women's Country,* present nonwhite characters in affirming ways (they are invisible in Sheri Tepper's novel, as are gays), *Woman on the Edge of Time* is the most aware of racial equality — and emphasizes it. There is a strong note in the novel of other-than-white cultures, and a strong validation and respect for all people. While all of the books so far have also made clear separation between the sexes, Marge

Piercy shows men and women as fully equal, fully autonomous, and sharing a common vision. Her characters are gay, straight and bisexual.

Connie's first psychic "visit" to Mouth of Mattapoisett surprises her. She expects an ultra-tech future but finds a simple village, and people she can love and understand.

> She saw...a river, little no account buildings, strange structures like long-legged birds with sails that turned in the wind, a few large terracotta and yellow buildings, none bigger than a supermarket of her day, an ordinary supermarket in any shopping plaza. The bird objects were the tallest things around and they were scarcely higher than some of the pine trees she could see. A few lumpy free-form structures overrun with green vines. No skyscrapers, no spaceports, no traffic jam in the sky. "You sure we went in the right direction? Into the future?"...
>
> Most buildings were small and randomly scattered among trees and shrubbery and gardens, put together of scavenged old wood, old bricks and stones and cement blocks. Many were wildly decorated and overgrown with vines. She saw bicycles and people on foot. Clothes were hanging on lines near a long building — shirts flapping on wash lines! In the distance beyond a blue dome cows were grazing, ordinary black and white and brown and white cows chewing ordinary grass past a stone fence. Intensive plots of vegetables began between the huts and stretched into the distance. On a raised bed nearby a dark-skinned old man was puttering around what looked like spinach plants....
>
> "What's on top? Some kind of skylights?"
>
> "Rainwater-holding and solar energy. Our housing is above ground because of seepage — water table's close to the surface. We're almost wetland but not quite, so it's alright to build here. I'll show you other villages, different...I guess, compared to your time, there's less to see and hear. That time I came down on the streets of Manhattan, I thought I'd go deaf!...In a way we could half envy you, such fat, wasteful thing-filled times!"[20]

Connie is disappointed at first in a future that is different from her expectations, but in reality it is an ideal place to live. No one is poor in Mouth of Mattapoisett or in any other colony of its time. Everyone has what they need, and objects are "borrowed" from a central library rather than owned. The settlement raises its own food and is self-sufficient, or nearly so, with other needed things traded for with other villages. The people have private living spaces, as they choose, but meals are taken communally. Children are given basic education in daycare centers, then choose what they wish to learn more of by applying to apprentice with someone in that expertise. Luciente is a plant geneticist, and her work and creativity are valued. Creativity is evident in every building, the creativity of the people who live there.

Mouth of Mattapoisett is a village designated as Native American, but the residents are of all races. Other villages are designated as Ashkenaze Jewish or African-American and embrace these cultures. The people are a mix of races in every village. While the settlements are low-tech and living in the Earthway, they are not without technology or science. There is electricity (solar and wind-fueled), flying air cars, and public transportation. All decisions on land and technology use are made in the public councils, based on the earth's needs and the good of all the people. Resources are scarce and must be conserved, the earth taken care of, but no one goes without their needs in this "harm none" culture.

Though medicine has become healing here, there is still hi-tech birth. Birth in fact, is not through women's bodies but from artificial wombs. Women do not give birth live any longer, a concession by them to making men and women equal. Instead, both men and women are given hormone injections to make them able to nurse a baby. Three parents together accept the new infant and raise the child until independence. At least two parents nurse the infant and all are active in childcare. The parents can be both sexes, and parents and child can be any race or racial mix. When a member of the colony dies, request is made at the brooder to begin a new life, and the child is presented to its parents nine months later.

There is another side to the utopia of Mattapoisett, both in the 1976 mental hospital that holds Connie and in the 2137 world. The members of Mouth of Mattapoisett are called away in rotation to fight for their freedom, and fight an alternative repressive reality. Connie in her daily life is also fighting the imprisonment, manipulation and repression that if continued would result in the negative alternative manifesting in 2137. It must be stopped in Connie's time. Luciente asks Connie why it took her civilization so long to begin fighting for freedom:

> "Free. Our ancestors said that was the most beautiful word in the language." Luciente stopped to beg a swallow of wine from White Oak, wearing a long white tunic slit up the sides and toting a jug of red wine. "Connie! Tell me why it took so long for you lugs to get started? Grasp, it seems sometimes like you would put up with anything, anything at all, and pay for it through the teeth. How come it took so long to get together and start fighting for what was yours? It's running easy to know smart looking backward, but it seems as if people fought hardest against those who had a little more than themselves or often a little less, instead of the lugs who got richer and richer."

> "We hate ourselves sometimes, Luciente, worse than we hate the rich. When did I ever meet a richie face to face? The closest I ever came to somebody with real power was when I was standing there in front of the judge who sentenced me. The people I've hated, the power they have is just power over *me*. Big deal, some power!..."[21]

The issues of negative or positive use of power, of greedy power-over versus "harm none" power-within are central issues of this book, and central issues in how we create a New Age and change the present. Marge Piercy's *Woman on the Edge of Time* paints very clear pictures of the alternatives, negative and positive, of this use of power. Her carefully thought-out utopia of

Mouth of Mattapoisett is a thought to bring into Be-ing for a women's New Age.

The newest of the six books in this chapter is Pamela Sargent's *Venus of Shadows* (Bantam Books, 1990). This utopia is on another planet, the earth colony of Venus, in a series of domed cities far in the future. The people are pioneers living in a highly technological culture, fragilely protected from a hostile outside environment, and involved in the politics of their own destiny and freedom. Earth has become a difficult place to live, overcrowded and ruled by the Arab bloc; permission to emigrate is hard to obtain and there is no return. To leave earth, or even to apply to leave, one must give up all possessions. On Venus, everyone works according to their skill, and work is the highest-valued quality. The settlers are building a new world, a New Age, and are determined to make it a positive one.

With the help of the Habbers, people living in constructed artificial worlds, Venus is transformed by hi-technology.

> Anwara, the space station..., circled Earth's sister-world in a high orbit. The shield called the Parasol, an umbrella of giant panels with a diameter as large as the planet's, hid Venus from the sun, enabling that world to cool. Frozen hydrogen had been siphoned off from distant Saturn and hurled toward Venus in tanks, where the hydrogen combined with free oxygen to form water. The atmosphere had been seeded with new straits of algae that fed on sulfuric acid and then expelled it as iron and copper sulfides.

> Venus' first settlements had been the Islands, constructed to float in the planet's upper atmosphere slightly to the north of the equator. Platforms built on rows of large metal cells filled with helium were covered with dirt and then enclosed in impermeable domes. On the surface, construction equipment guided by engineers on these Islands had erected three metallic pyramids housing gravitational pulse engines...The planet,...had begun to turn more rapidly.

> Domed settlements were rising in the Maxwell Mountains of the northern landmass known as Ishtar Terra. The people who called themselves Cytherians, after the Mediterranean island of Cytherea where the goddess Aphrodite had once been worshipped, were living on the world that bore one of that ancient deity's names. [22]

Risa Liangharad is the protagonist of *Venus of Shadows,* a woman born in the Venus colonies. The book follows her life and that of her family on the domed surface of a new world. In the early days, the Oberg dome is a quiet, safe place to live. Population is small and the rules are simple and few. No one locks their doors, and everyone knows everyone else in the settlement. Most of the houses contain greenhouses where food is grown.

> The main dome was five kilometers in diameter. Oberg's able-bodied citizens moved through their settlement on foot whenever possible; the walking exercised their bodies and kept the domes free of too many roadways and vehicles. She moved away from the houses and followed a small creek to the west....

She came to another sloping ramp and went up into the west done; ahead of her, a curved path led away from the main road toward a group of dwellings and small greenhouses. Trees hid some of the houses from view, and much of the flat land surrounding them was covered by shrubs and high grass....

Their home now was a one-story building with two small rectangular wings; it sat among slender trees about thirty meters from the entrance to the tunnel. Chen had built most of the house himself, from materials given to him. She and her father had been among the earliest settlers, those who had come to Oberg sixteen years ago when only its main dome was ready....Now ten domed settlements sat on plateaus among the Maxwell Mountains, and thousands of settlers had followed Risa and Chen to the surface.[23]

Life on Oberg is based on work, the work necessary to maintain life in a hostile environment. Risa explains to a new arrival:

Oberg has three domes with settlers and a fourth being made ready for more, but everything we build has to be maintained. Every Cytherian, including the children, learns at least one skill that can aid the community as a whole. Every action has to be measured against certain limits. You may, like many immigrants, be wondering why we don't clear more land for houses here so that you don't have to live in a tent, but those trees and plants help to maintain our oxygen level inside. We have a small lake in the center of the west dome, but it isn't there just for our pleasure — it and the streams you'll see keep the air from becoming too arid....Much of the work we do involves simply trying to hang onto what we have....

You'll have a chance to learn more than you would have (on earth) because that may make you a wiser and more valuable citizen. You'll be rewarded for good work, not just with credit, but with the respect others will give you if you earn it. You can make your life what you want it to be, instead of letting the Mukhtars and their representatives decide that for you. You can know that your children have a dream and a purpose. That's our real freedom — to know that our children will have more."[24]

There is a hi-tech but simple medical system on Oberg, with rejuv shots to prevent the worst effects of aging. When the injury or dis-ease is beyond what can be managed in the dome, the patient is sent to one of the Islands. If the problem is more than the Islands can heal, nothing more can be done; they are not sent off planet. There is an emphasis on prevention, but it is hi-tech medicine and not healing, on Venus.

The Oberg Council runs the settlement, and is comprised of five elected members. They mediate disputes and try to prevent appeals from going higher up, to the Islands or earth administrators. After two years' residency every adult citizen has voting privileges. Town meetings can be called, and one is described, but the democracy is not generally participatory. Marriages are taken seriously, and homosexuality becomes severely repressed, for awhile, until the people revolt and regain their freedom.

An interesting turn-about on Venus is religion. While eastern religions seem represented — a Muslim mosque and Buddhist temple are mentioned — a fundamentalist Goddess cult develops in the domes that gains in power despite the non-acceptance of most settlers. The group is called the Ishtars. They describe the Goddess as the planet herself who they must supplicate and merge with. Part of their "merging" is in group eroticism that is frowned upon by the generally conservative residents. A large focus of the book is in the increasing take-over of this power-over group and its eventual downfall. In actuality, the Goddess religion has never been either fundamentalist or power-based, and perhaps this is a warning for women to maintain the balance. The worldwide takeover of any religion, even the Goddess, is a negative thing. The power dynamics and politics are destribed in detail in this huge, six hundred-page book.

Whatever the internal workings, survival is the first agenda on Venus, and second to that is autonomy from the Mukhtars of earth. The view through the dome is enough of a reminder:

> Lightning flashed for a second, revealing black, barren mountains so high he could see no valleys below. Droplets of acid rain glistened on the impermeable dome surface; in the distance, on another rocky shelf, he glimpsed the faint glow of another settlement. He had assumed that one reason for using a transparent material for these domes was to keep the settlers from feeling too closed in, but this sight made him uneasy and aware of how precarious humanity's hold on this world was.

> "We have our limits, Nikolai," Risa continued, "The Mukhtars may often be far from our thoughts, but we aren't exactly free to do as we please. Out there, the atmospheric pressure would crush you, the heat would boil your blood, and the rain would eat away at your bones. Every time there's a quake, the dome's able to withstand it, but we have to send out our diggers and crawlers to make sure rocks are cleared away, and then see that none of our installations have been damaged." [25]

The domed cities described in future life progressions were questionable but they were seen as future ways of life after the earth changes. Life on other planets and colonies in space were shown as alternatives for the future. The domed colonies of Venus in *Venus of Shadows* describe this reality, and human attempts to create a New Age utopia against the harshness of survival in space. Perhaps the least positive of women's visions, it is a choice of realities.

The women's novels of this chapter were chosen for their alternatives. All of them describe a world that was seen as a future possibility in the channelings or future life progressions. By reading women's visions of a New Age, the future comes to life, and if the vision seems undesirable, women can change it. From the material of this book, women can think into Be-ing these visions or others, whatever appears the best. There are a multitude of choices to consider.

In the past of women's herstory was another civilization, now lost to us, with lessons for today. In the present is a time of change — emotional, geophysical, international and personal — a time for women to define what needs changing and begin it. It is also a time of defining what the future could be in the best of all possible worlds — and to begin putting that into effect. The future is what women make it, what we are defining, devising and developing now. By women's power of thought and creation, we can make a better world. With all possible futures and the women's New Age before us, let us now begin.

Full Moon in Aquarius,
August 6, 1990

Notes

1. Mary Summer Rain, *Phoenix Rising*, front page.
2. Sunlight, *Womonseed* (Little River, CA: Tough Dove Books, 1986), pp. 4-5.
3. *Ibid.*, p. 6.
4. *Ibid.*, p. 107.
5. *Ibid.*, p. 127.
6. *Ibid.*, pp. 150-151.
7. Sally Miller Gearhart, *The Wanderground: Stories of the Hill Women* (Watertown, MA: Persephone Press, 1978), p. 63.
8. *Ibid.*, pp. 2-3.
9. *Ibid.*, p. 2.
10. *Ibid.*, p. 13.
11. *Ibid.*, p. 106.
12. *Ibid.*, p. 158.
13. *Ibid.*, p. 128.
14. Suzy McKee Charnas, *Motherlines* (New York: Berkeley Books, 1978), pp. 59-61.
15. *Ibid.*, p. 190.
16. Sheri S. Tepper, *The Gate to Women's Country* (New York: Bantam Books, 1989), pp. 301-302.
17. *Ibid.*, p. 61.
18. *Ibid.*, p. 258.
19. *Ibid.*, p. 288.
20. Marge Piercy, *Woman on the Edge of Time* (New York: Fawcett Books, 1976), pp. 68-69.
21. *Ibid.*, p. 177.
22. Pamela Sargent, *Venus of Shadows* (New York: Bantam Books, 1990), pp. 5-6.
23. *Ibid.*, pp. 108-109.
24. *Ibid.*, p. 122.
25. *Ibid.*, pp. 121-122.

The Channeling Sessions

Mari Aleva — February 5, 1990

The following is from a tape with Mari Aleva channeling the energies of Helm and Delta Four.

Diane: I'm looking for information for my book on earth changes particularly geared toward and for women on a couple of particular subjects. One is Atlantis, women's role in Atlantis, maybe what happened in Atlantis as far as the destruction of Atlantis. On the creation of the soul, why we're in bodies, why we're here on earth. On the earth changes, on what is coming in the earth changes, on the earth changes both as a geophysical thing and as the evolution of the human consciousness. I'm looking for earth changes material that's geared toward people, women particularly, whatever you and Helm choose to address.

Mari: This is Mari who is alive and aware on the back burner. If something is unclear to me, Helm's liable to ask questions like "Is this valid for you?" or "Does this make sense?" It's not her, it's me pushing because of nervous conditions here. There may be other guides coming through because she was talking about that. They're lined up to share.

Before we started, I called in the directions and the energies from each direction that I felt personally were needed for myself and for the assistance in this channeling. I also protected the area with incense and asked that all the energies here be for the highest and greatest good for myself, for Diane, for Caridwyn and for anyone else concerned with or connected with this taping and the book. We want only energies that are for the highest and greatest good of all. And I do feel there are more energies here than my own guides. I feel there are guides of yours...

Diane: I feel a lot here.

Mari: Guides of Caridwyn's and there are guides that are coming just to hang out.

Diane: Welcome.

Mari: So if either of you pick up anything either kinesthetically or just feel something or if either of you hear or sense something, or see visions, it would be, I would say, an advantage for you to share also. And anyone listening to this tape or reading the book that comes out, know that your own experience, what you are picturing as you read the words, what you're hearing from within as you read or hear the words, and what your personal feelings are, is very very valid, even if it does not coincide with what is being channeled here. It may open some of your own internal doors. As we all know, three or four people can go and see the same movie and each one will see something different. So there's no right or wrong, there's only experience and what you want to get out of it. So with all that...

Helm: Ha, this is Helm. Greetings.

Others: Greetings, blessings.

Helm: And greetings over here, too. Perhaps we are too loud.

Diane: No, that's all right.

Helm: We can smell your incense here. Smell is very important, not only to set mood, but for movement of your cellular content within your body. Aroma therapy does assist and you ask about things from ancient time, like Atlantis.

Diane: Yes.

Helm: Aromatherapy was very popular there and these memories are within your own bodies. There are many times that people here smell a certain aura, aroma, and it smells very strangely familiar and yet it is new. It brings back memories, cellular memories. You see we believe what we teach is the body itself has memories beyond your own soul self body. What we are speaking of is the DNA from your parentage all the way down from your parents and their great grandparents. You have all of these memories that go down what you call your blood line.

There are also what you would call spiritual connections which are the essence of your core soul before and after it combines with your body and what combines it would be mental attitudes and emotional input. The spiritual part transcends way beyond this earth, for it is our belief that there have been people come to earth that have not evolved from the amoeba. They have been sent here in what you would call a space ship. There are many things connected with what you call space ships, having to do with interdimensional travels and bringing their vibrations in connection with the earth so that they can be here to live and to colonize.

Diane: People came to earth initially from other dimensions? Is that what you're saying?

Helm: There are people here, yes.

Diane: Okay. Was earth settled that way? Did humans come here initially that way?

Helm: Certain parts of the earth were settled this way. Some of this, what you call Atlantis, is before earth change here. People descended from other place, not evolve from here. Women in ancient civilizations, that Helm knows, has experience with, working with the crystalline elements, the women were not only the nurturers, but the providers, as well. They were the gatherers and the teachers, for when the childrens were born to the women, the women would be the ones to teach and to bring up the family as the number one priority of the family. And then they say, well, who will take care of the children? You have extended families, sisters. You see, the men would do what their bodies do, is the strong things, to build. In the old days here on earth you call that slaves. Only we are not saying men were slaves per se. But they would build. The women would design, they would build. And there was also when there were invasions, the men would stand guard. There are many ancient traditions who believed in the spiritual realities and they would train to protect against this.

Diane: So the women were basically dominant and the men were protectors?

Helm: There have been many cultures, the culture that we can share with you does have what you are saying. The women were leaders in those communities.

Diane: And that was in the communities that we call Atlantis, but were other dimensions rather than the earth plane as we know it?

Helm: We see there were some in male bodies who were very wise ones and were what you would call in this time, not convent, but in Tibet they have them, and what are they called?

Diane: Monasteries.

Helm: Monasteries, type of thing, where they go in there and they work with wisdoms and they work their own inner magic. The men would do this and the women would do this, too.

Diane: Was it women then that abused the technology? And caused the end of it all?

Helm: The technology is another story. There were not only the people from one country involved here. There were others wanting what some people have, other one want to have it, and there would be destruction rather than give it away. Does this sound familiar?

Diane: All too familiar.

Helm: There were eruptions of lands. There were shifts of land, from pull apart from different...

Diane: A pole shift?

Helm: Poles would do, magnetics, has to do with sky and with your gravitational pulls of energies bypass around and it would combust. We have a barrier here with language.

Caridwyn: Helm, does this relate at all to the land break between Africa and South America where the pieces fit together if you move them together, but they began to drift apart after a major rupture?

Helm: You must understand. I want to tell Madelein (Mari) something. First for you. Understand of land erosion, lot of erosion here.

Diane: Wearing away?

Helm: Now to tell, I want to speak with Madelein, here. She say how can land float? Let us explain, honey. You have your core, and you have a liquid mass here. There is metallics in the liquid mass and there is your polars. Then as it cool you have your crust. And as crust splits apart, it is still connected to its core and you have what you call water which is liquid and can move. The land as things move can move when it is connected to a core. Do you understand?

Diane: Yes.

Caridwyn: These are the tectonic plates that they talk about.

Helm: Well, Madelein, here, does not understand. This is only way we can explain, connected and they can move.

Caridwyn: It's as if you had the center of a wheel with the spokes moving out and if some of the spokes start moving apart, they're still connected to the center though there may be wider spaces.

Helm: Absolutely. That is right. You see, Madelein think that you talk about island, that is like a piece of ice that float on top of water.

Caridwyn: No, it goes down deep....

Diane: So, this was the pole shift that destroyed Atlantis.

Helm: As the pole shift, think of her wheel here and here and as things move they do wiggle. Ah, and if it move here, it is still connected and the water in between erodes. The cultures come back and forth around from once walk on land and they say, how does this happen? There is a ship underneath...

Ah, hello, puppy, how are you?

Diane: Can I let her go?

Helm: You can let her go. She does not eat our crystal here.

Diane: No, she won't bother it.

Helm: I want to share something with you. Your Reiki can help her hips. Have you tried this?

Diane: A little bit. Okay, I'll do more then. Thank you.

Helm: When you put hands...doggies have five chakras, rather than seven.

Diane: Oh good, I thought I maybe wasn't finding enough of them. That's what I found was five, too.

Helm: I wanted to confirm that. Our kitties have five.

Diane: All animals?

Helm: Not all. These and the majority of them you will have five, not seven. Major chakra areas.

Diane: Okay, is there anything else I can do for her?

Helm: It would assist if she had straw in her house for soft on her. The floor is sometimes cold, is not good for bone.

Diane: Okay, how about a bath mat or something like that?

Helm: Could do.

Diane: She doesn't go in there that much. She gets to sleep on my bed if she chooses.

Helm: Well, if you ask, we say.

Diane: Thank you.

Helm: She's a very good dog. Has been very loyal to you.

Diane: For sure, for sure.

Helm: We like doggies.

Diane: Me, too.

Helm: We like animals and animals can tell you things. As you connect with her and her mama. Her other one is around, you know, you talk to her.

Diane: Yes. Oh, yes.

Helm: These doggies can be guides. They say animal guides, well they tell about the Native American ones. The spirit of animals can be connected to your own core soul or core soul group as Madelein explained to you earlier today. Energy is energy, you attract the type of energies that you are putting out signals for. It is like television. You turn a channel on and if it does not attract you, change the channel. Animals are very close to earth. Some of them have also been shipped here, not of earth.

Diane: Ah, okay, I understand. Dusty's mother, I have always felt, was a lot more than a dog.

Helm: Well sometimes energies choose to come as a dog, as a pet, migrate toward a certain person for experience of soul for you have owned this dog in other lifetimes, have you not?

Diane: Yes. I don't own them. I live with them.

Helm: Yes, have shared abode. Caridwyn, you are being very quiet.

Caridwyn: I've said enough. It's nice to be with you, Helm.

Helm: Well, it is nice to be with you, too. You have quite a good talent. As you practice with your typing and your channels, you will go far. You are very good at what you are doing. For when energies are coming through, they work with your own mental attitudes, mental body, mental auric field, to be able to communicate through you. Let us say you have all died here, you are all no bodies, and you want to come back and guide our grandchildren or our great grand nieces or whoever you want to guide, and they are speaking a different language, how would you contact them?

Diane: Through the senses?

Helm: That is one way.

Diane: It would seem that the energy would translate into whatever language it needed to be in.

Helm: And the translator is the receiver's library.

Diane: Okay. How did souls come to be in bodies anyway to begin with?

Helm: Manifestations.

Diane: Why did they choose to do that? Why are we here on the earth plane in bodies doing what we are doing? That is a lot to ask.

Helm: That is a good question. When you have a purpose in mind, a calling, a reason to manifest a reality, sort of like go to schoolroom, you vibrate to the energies of earth and you manifest whatever. This is for here. There are many other places where manifestations occur and these manifestations are in many forms. Some are translucent, some as rock. If you are a void, only energy, and you want to use form and feel senses, then you like jump into a page of a book. You vibrate to the energies and manifest to where you want to go.

Diane: Why did people choose to manifest in bodies? And why did people choose to be on earth? I'm asking too much?

Helm: Depends on people. Each one will have their own reasons. Madelein chose to come here to assist in the earth vibration change. You see, all of these are connected. People say, "Helm, why are you coming through and talking, raising consciousness?" Well, the reason why is cause when you turn on your light, other people around catch the glow. You are in a room and you teach, it heightens all energies. When Helm come through it allows your lights to brighten. Helm like to brighten. And what you share, helps us brighten. It is to increase in light knowledge. And we get brownie points from our own guides and teachers. You see, we have guides, too. We are friends. There is not only that we are all knowing all. People ask Helm all these questions like Helm would know everything. Figure you have no body and you are contacting somebody here and they ask you how to speak French. How would you answer?

Diane: I don't know how I'd answer.

Helm: So there are some questions your guides may not be able to answer. When you allow them not to answer, then you are getting some truth here.

Diane: Will they tell us, though?

Helm: When you push for the truth, they will tell you "I do not know the answer." When you push sometime maybe one of them will make a nice story that sound good [discernment needed? CT]. And that is all right for those stories come from somewhere. You asked us a question. We talked right on top of you. Go on ask it again.

Diane: I don't remember the exact words, but I was asking if that's what's been happening to me personally in the last few weeks.

Helm: Yes, yes.

Diane: Okay. Why would not in that case the guides say "I don't know," or say "Don't ask any more?" or make some kind of an indication.

Helm: I cannot speak for other ones' guides. I can only speak for myself. There have been times I have told Madelein and others. "I'm sorry, we do not know." And sometimes we check out to let another guide come through that knows more about a particular subject. Like we can come through with a lot of things about art and we can communicate through color with Madelein because of her use and knowledge of color as an artist.

[Telephone.] Breathe life into the body. It is like coming to earth to enter more of these questions. Many reasons people had to come to earth. Sometime for a personal endeavor, sometimes for a quest of knowledge to learn and to experience something here. Sometime it is to connect with others in your own core group for a group purpose. Sort of like a class project, if you will, through experience.

Diane: Was earth settled for that reason? To be this type of a laboratory?

Helm: Not necessarily. Energy is energy. All energy is in its own form life. There are many dimensions, many kinds of things, because energies always change. Earth forms from energy, combustion. They call it "Big Bang." We don't like that expression, but it is all right. And it was a piece of this part that went and cooled down to where it is now. This is gone through the books. Everybody here has read this books.

Diane: The one that really interested me was someone named Sitchin who writes about earth being the 12th planet and having been a part of a planet that was destroyed by another planet passing through, moved out of orbit as a fragment to become the earth, that the earth was originally called Tiamat, or what the people in early times called Tiamat.

Helm: You will have to go quite deep and get into these other informations. All we know is a picture, names we do not know, of a thing and a part of it is what you may call your earth. And your sun, too. As the sun cool, it will change your temperatures. Long time from here, millions of years. Worry about it for your great, great, great grandchildren, that is, if the earth still decides to be here.

Diane: What about our own children's future on this planet? As humans we are not doing well by this planet.

Helm: There are many choices for future, and there is a mass change occurring right now, starting.

Diane: Can you talk about some of that and some of what's coming? As it appears right now, I realize that it's a changing thing.

Delta Four: These are the energies of Delta, what has been known as Delta. Greetings.

Diane: Welcome! Thank you for coming.

Caridwyn: Greetings, Delta.

Delta: Welcome. We are coming through to assist Helm in her delivery. The energies known as what we call Delta Four are from a different dimension. That dimension includes the dimension that you are familiar with, which is the third. Each time the earth passes another moment, it impresses upon time an alternative existence that continues. When you ask your questions regarding the earth energies, where they descend from, why they are here, what the form is connected to, and where it is going, yes, it is a mouthful, but we can assist in answering. It is thought form. [Diane: Thank you.] The collective consciousness which you may call the All There Is, the Goddess energies, God energies, all energy that has ever been has many forms of energy. As a certain amount of connectedness grows, they form thoughts which creates mass. This mass then becomes a vibration of reality of combined energies. If in fact there was a planet that was made by these energies in a certain time and space that was then dismantled through combustion explosions causing flame and fire and transformation, and perhaps the cooling process of the part that you may think is earth at this time, these are possibilities. And then where do the inhabitants come from? How has nature evolved from flame to flourishing trees? As we all know, the change is inevitable. Change takes many forms. Energy moves forward, and energy moves backwards. And when it moves, it moves at a different vibrational rate, different time frames, and it is a very, very intricate and deep subject that entails a lot of studying, a lot of pages of your words.

The part that we feel you are interested in at this time is the group thought of earth here now. There are many dynamic group thought processes. Some are what you would base negative, some positive, and a lot in between. At one level we have your warriors, your gun crazies, whatever you want to label them. And at the other end you have your peace makers and you have everything in between including the so-called churches that bring the light of the Christ within and they harbor guns. Thought forms are creating new energies and the light communities of the new age expand, look for the other side of the coin to be there. It is similar to the angelics opposing the demons. Does this answer your question?

Diane: It answers a lot of questions, yes. That even though change may bring destruction, the destruction is of the old and something new will come that could be a lot better. Am I interpreting, understanding that correctly?

Delta: Destruction occurs in many ways. Some destruction, what is termed destruction, is actually a rebirthing. As you destroy the old skin, the snake continues to live.

Diane: What about all the changes inside individual people that are happening right now, the type of destruction defined as death and rebirth that I am referring to? I see that as a major part of what earth changes means is the changes in consciousness in individuals.

Delta: As the consciousness changes, your earth evolves in that direction.

Diane: Is it possible that people's changing, the consciousness changing of people, can prevent the physical changes that are prophesied, that were prophesied in the past?

Delta: Changes always occur. You can ask your animal life, for they hold many answers to the changings of earth. Earth is filled with poisons made from humankind. We will leave but return again soon. Farewell.

Diane: Thank you.

Helm: Madelein, you should not do this. We'll bring Delta back again. We have this one here. She did that one. Let us take a break. Hold off your thing here (tape recorder).

Diane: Turn it off? Okay.

[Break]

Delta: Caridwyn, we appreciate the analogy. Energy is energy is what Helm has been teaching. The earth energies of nature are there to nurture and to assist humankind in their growth, animal kind in their growth. It is interesting that you picked up on the demonology and the angelics, for they are both thought forms. It has been written, "As a man thinketh, so is he..."

Diane: What you think, you create?

Delta: Similar to that and there are words written. We are not picking it from the directory of the channel. What has been written is what the reality becomes (for those who read it). The thought creates forms. The earth energies are very concerned about thought forms of humankind, for the survival of the earth energies are dependant upon the energy thought forms that humankind places upon the earth. Think of the poisons and the death of the many trees causing imbalance in the air. When you grow trees, it will clean the air, basis of your Greenpeace. Your larger cities will find more illnesses and poisons in the air.

Diane: They already are increasing.

Delta: These also have things to do with the mind and the creating of the energy in a different form. What we were saying in Dianna's basement in Michigan, the extension of that about the creating of energy change and humankind changing and evolving along with the inventions that are destroying the earth. It's like the bugs that get used to the bug sprays and pretty soon it does not work.

[People will keep inventing worse things to kill each other. — Mari 2/25/90]

Delta: People are changing in the same ways, and new diseases are coming.

Diane: I understand why they're coming, but what can we do to change this, other than to plant trees and to raise people's awareness about caretaking of the earth and about positive thought?

Delta: Thought form is the utmost importance. Do be prepared though, to realize the opposition.

Diane: What about all the — and I don't know if you would class this as negative thought or not — but since the 1987 harmonic convergence, people have had an onslaught of pain in their lives, and I realize that's a part of change and a part of rebirth. Is that also negative thought, is that pain negative thought, or do you class negative thought the penchant for war and money making and pollution rather than the pain of well-meaning individual people who are suffering right now?

Delta: Negative thought would be those energies that you no longer want in your life. And they are different for different people. We are going to bring

Helm in to assist on answering the next one.

Diane: Thank you. Blessings and thank you.

Delta: Blessing to you and to you. Farewell,

Caridwyn: Farewell, Delta, and thank you.

Helm: Well, I come back.

Diane: Welcome, Helm.

Helm: Greetings to you.

Diane: Greetings. Thank you for being here.

Helm: We mention the harmonic convergence. We speak about these another time. Madelein is much more interested in that one and she's comfortable with Helm. Delta is still relatively new energies and difficult sometime to hold for long periods.

Diane: Do you want to talk about the harmonic convergence?

Helm: Harmonic convergence is a very interesting thing. It involves the planetaries, the planets, the different poles and how things were in a precise time, somewhat like when Jesay was born, Jesus you call. There were energy changes that occurred during this time. Many doors have been opened. You're speaking of much negative. Ah, but know as these things, the other side as Delta mentioned, you have ones that are growing.

Diane: So it's the release of negativity?

Helm: Many people you'll notice new opening. They come to you and they say, well a couple of years ago they do not know harmonic convergence, and they go couple of years ago I start getting into this thing or then the last year most of them will say since 1987 or 88, 89 they have been getting interested in these things. Because of the energy change.

Diane: Okay. More and more people.

Helm: And another thing, you can put this one in your book, this is another one going to confirm if it is not already there. We do not know. The pole in the earth have changed a fraction of inches, very few. As that occurs, the magnetic things in there change and that does cause some disruptions around your earth.

Diane: Is there any connection between that and the hole in the ozone layer, or is that entirely human caused?

Helm: The hole in the ozone is aggravated from the sun burning up through here.

Diane: Because of human misuse?

Helm: Yes, there are things that are natural, that are not natural. You'll find these holes where there are not many trees.

Diane: Yes, right, okay. People are making such immense changes. That ultimately is positive?

Helm: Ultimately we think it is.

Diane: Okay. But that's also creating a great deal of negative energy on a personal level. Is it affecting on a planetary level?

Helm: Oh, for some people it is making a negative. And for some people a positive. And these are judgments. What is positive for one person, Madelein would think it would be good to lose twenty pounds, thirty pounds off her body. Take an anorexic person and it would not be positive.

Diane: But on individual terms, people are changing and that's because of the change of energy? They are growing.

Helm: Yes, what you would call universal world change. The gathered thought of the planet, it is like these crystals. As they formed, crystal stone, as they formed they also pick up what is around them in energies like the life forms. Their energies, you can pick up where it comes from, with your third eye here, for the energies are there similar to your psychics who touch somebody's metal ring and they can tell about this person, who it was and see pictures. And these things are in like crystals things from the whole culture. So your greatgrandchildren's cultures can pick something from this era and know what is happening.

Diane: Ah, okay. I see on a planetary level as far as nations go a lot of very, very positive things happening. The Berlin wall coming down, changes in Africa, changes in Eastern Europe. But on an individual level it doesn't seem so positive or is it because the changes are not complete yet?

Helm: Changes are happening. The last twenty years of this millennium has been predicted of much change. There was for Madelein a war predicted that did not occur.

Diane: So that as people are changing some of the negative things don't have to happen?

Helm: Absolutely.

Diane: Okay.

Helm: For she has predicted Ronald Reagan would become president, also we figure he die in office. They shot him; he did not die. These things were predicted when we live in California in the 1960s. He was governor. So some things will come and some will not. A war was predicted where we here would be involved. It did not occur.

Diane: There are prophesies of a war with China. Is that a necessary thing, or can that be prevented?

Helm: We are not knowing. We were going to share about things happening that are opening up. Also watch that the thing open up here, and you don't know what the other hand is doing, 'cause you are looking at this wall come down here (Berlin Wall). Always know that there is another hand somewhere. You mention China. How about India? What is happening with India?

Diane: I don't honestly know. They're developing nuclear power, which is kind of a concern.

Helm: You bet your bottom dollar they are.

Diane: Okay, and that's a threat.

Helm: The ones you don't always hear from, the quiet ones, might just surprise people. We are not wanting to think war. The more people who think war bring it into possibility. There are wars going on all over these countries. Some take in all entire countries, some entire cultures, some are in religions, some in people's own where they work and some in their homes.

Diane: Will that change for the positive?

Helm: It is individual. It goes from large world to even inside of people's homes and from there into their minds. When they change inside their minds and they open and feel better in their homes, then the communities get better and the countries and then the world. Starts from here (the heart).

Diane: I'm finding many, many, many women are working on the issues of having been abused or incested as children. How does that connect?

Helm: That is very sad. It (abuse and incest) has been going on for a long time. It started with people owning other people. Well, when someone says this is a sin, there are some people that are very attracted to what they call a sin. Do you know there are cultures where everyone sleep in the same bed, and if someone want to stroke a person to make them feel better, they do it. The children sleep in the bed where the parents are creating new babies.

Diane: But there's a difference between that and violence and invasion.

Helm: It is in the mind. And you see when someone forces another because they have been taught that this is wrong, then it is for them wrong. If they in fact have been taught this is part of the culture, that it is all right, then it is all right for that culture.

Diane: What I am saying is, yes, this has been going on for a very long time, but it's now that people are dealing with it.

Helm: It definitely has. The media has brought it to the attention of nationwide coverage.

Diane: Has it really come through the media?

Helm: It has been coming through. You go on your talk shows, they tell you about these things. And they are being brought out, and since these are happening, you have your homes (shelters and safe houses) that embrace victims and people that are assisting. And you also have the other side of the coin, people who do not believe that this is happening, and you have the ones that are causing it that want to hide the facts and so they say it does not happen.

Diane: How are women dealing with the emotional negativity after-effects of this issue? How is that change of consciousness in them connected with the idea of a change of group consciousness as an earth change idea?

Helm: On an individual level. Each person deals within their own individuality, and then it will grow out. It does not occur where it come from "out there" and come in.

Diane: So it's part of the individual changes that eventually will make the mass changes that will change the consciousness of the planet?

Helm: That is correct.

Diane: And it's a healing.

Helm: Of course it is a healing. Yes, it is a healing. And when you understand these things and put them in your book for others to read, many will know.

Diane: That's why I wanted to ask, because it is such an intense issue for so many people.

Helm: Did you realize that in the ancient cultures when the women were running things, if a man was caught doing these things against the will, they were castrated?

Diane: Sounds like a good idea.

Helm: Well, if you were a man, you wouldn't agree. But there were cultures where this did occur, and it was done barbarically.

Diane: I'll bet in those days, sure. Culture, human culture has changed so much, it seems like you're telling me that back further in the past women were a major dominating force. And I'm also understanding that what part of the earth changes means is it's been named by some as an Age of Women. So what's coming for women as human consciousness changes, as the earth changes and we solve or not solve some of the problems of the planet and move into what comes next?

Helm: They are gaining in strength, and the men are gaining in fear. And out of fear comes some pretty erratic behavior. Women are gaining in their equalities, they have been.

Diane: Will they be able to hold on to that and increase that?

Helm: There are many levels. The men are feeling fear. When ones feel fear they do not always think wisely.

Diane: So further oppression can come from that. I think we see that in the politics.

Helm: That is correct.

Diane: Okay. What can women do to hold onto their gains?

Helm: Support each other. Support each other and open the mind and think positively and think in light, what they call good stuff. It is much, an ounce of love will take care of a pound of hate. But you need to project that ounce of love. It is much stronger than hate.

Diane: I also see some men changing, and some men becoming gentle and opening their own feminine sides.

Helm: That is correct. Women are becoming stronger. Men are becoming softer. And it is to more of an equality, like we say, and of course there are the fringes, your rednecks and your high-heeled Barbie dolls at the ends here. And in the center different grades, in here would be your androgynous whatever and this grey area here is growing and it is more balanced in the center here where you have a balance of your femininities and your masculinities or your strengths on both sides. These are coming back. There were ancient times when women were more, there was the culture where we had, ah, don't know the words, but we had like a big board with different lights and switches on it, matrices and different things and women would run this and some of the men would, too. Where they did not care about which sex they were, but who was astute in the head, who was good in the head. And the ones who wanted to learn and were good at what they do, they would put them in a position that was best for that person regardless of their color or whatever.

Diane: What did the matrix boards do? What were they for?

Helm: Oh, powers of the crystals, of course. And they would light and they would do things. There was energy boards.

Diane: And this was in what we would call Atlantis?

Helm: Oh, yes, it was. They lived a lot of time in caves. They were underground.

Diane: I had read some of that and wondered, yes. They were an alternate dimension from earth rather than being on earth?

Helm: Both.

Diane: Ah, okay.

Helm: There are other dimensions, alternate dimensions, that are still alive. That is why we are connecting with them. You have visited.

Diane: Yes.

Helm: And there is quite real. You are perhaps their dream. We are going to leave soon. It is getting late. One more question.

Diane: What advice can you give to women through these changes?

Helm: Well, how about we do a personal one for you?

Diane: All right.

Helm: Think about what you would want most in your life for you. If no other people were around. They will be there for you. But know that you find your love and your happiness within your own self first. Radiate this and it will attract those to you that you are searching for. There are good things out there for all of us.

If you want to think in one side and the other side of the coin, and you've been going through what you call your shit side here, it is time to turn it over. That the other side exists as well or the bullshit you have been going through would not be real. When a door closes, another one of equal or better value is opening. You've heard this one before. It is true.

Diane: Right. I think I'm at the turning point right now.

Helm: You are, and your books are doing very well, and you will be writing three best sellers very soon. Five years from now we will have this talk and you will say, Helm, you are right.

Diane: All right. Thank you.

Helm: You are welcome. Au revoir.

Diane: Au revoir.

[Transcribed February 24 and 25, 1990, and edited October 6, 1990, by Caridwyn Aleva.]

The Channeling Sessions

Laurel Steinhice — June 3, 1990

Note: In this session Laurel Steinhice began by channeling the Earth Mother, who she says is quite near and now being received by a number of channelers. She then switched to Edgar Cayce, the entity she most often channels. She is the only living channeler through whom he speaks publicly at this level.

Mary: ...And this relates directly to the earth healing and the transition which is what you wish to speak of. When you heal yourself and assist others with their self-healing, you heal the earth. For the earth is one with everyone and every creature. All are part of the earth. Those that creep and crawl and go upon many legs, those that fly in the air, those that go upon four legs or two, are the Earth Mother's children. All are a part of the earth. The healing of any one of these is healing that contributes to the whole.

Diane: There's so much to do, and I try so hard. Does it really make a difference?

Mary: But you do (make a difference). That is the message we bring. *It does matter*. The simple loving of the earth is healing. When you love the earth you bring energy for her healing. There is interconnectedness between earth energy and personal energy, and use of that energy is what the transition is all about.

We will speak again in this direction, but for the moment we will let Edgar through to talk about earth changes.

Diane: May I ask in what name you are speaking to me?

Mary: I am Mary. I am also the Goddess, known by many names. You are familiar with my energies....

Edgar: We are Edgar again. It is my persona that does most of the talking through this channel because my vibrational frequency is nearest to her own. And it is therefore the fastest and the easiest.

We would speak of earth changes. When I came to this channel not quite two years ago, I specifically said it was my first appearance on the earth plane since my death crossing transition in 1945 and announced that I had come to assist with certain projects and needs. I am available to serve as consultant for spirit and guide teams who may wish my involvement. I come not as a god-figure, for I am certainly no such thing, but as the voice of experience, for I have been the channel as well as the guide. And I am now, as I was then, very interested in personal healing, earth healing, and projection of probabilities, which is called prophecy. I assist many seers and healers, many channelers. One of the fringe benefits of being dead is being able to be in many places at one time, and participate on many levels simultaneously.

I am Edgar Cayce, who was called the Sleeping Prophet. I am many others, also many past incarnations, and I am many fellow selves. We give the name the Crystal Circle to ourselves merely as a way of designating that we are connected, allied with each other, sharing the search for enlightenment, for this is endless. It does not stop with moving from earth to spirit plane. It is the continual pathway of enlightenment, of oneness with the universe. We share the path. We share an interest in metaphysical exploration, we share a desire to serve, we affirm the positive nature of the universe, the love of the Creator, Father-Mother, Mother-Father God, the All That Is. No matter what name you call it by, it is the one creative voice of the universe. Sometimes we just say "The Force." It is a wonderful understanding that anyone can grasp.

We affirm the connectedness of one spirit and another, whether incarnate or discarnate. And we practice cross-plane communication, speaking to those who are incarnate and discarnate.

Diane: Like those of us who run around in the astral!

Edgar: Oh, there are many such! This planet is changing. One does not need to be psychic to see that. Changes around you everywhere, everyday. And it is changing so fast! The whole program is changing so fast. We bring updates from time to time.

By loving the earth, you help to heal it. Personal caring makes more difference than people realize. Let us give you some examples of this. In 1988 there was a great hurricane in the Gulf of Mexico. And everyone said, "This is the worst storm in a hundred years." Perhaps you meditated or prayed, or simply sat before your television set and said, "Oh, we hope this doesn't go ashore where it will do the most damage."

This caring, the positive energy of persons of good will everywhere, whether incarnate or discarnate, combined to push this hurricane away from where it would have caused the most harm. This is your success, and that of many. We invite you to know that you are very much a part of this and share the joy, share the satisfaction of a job well done.

In the same way, the negative energy which would come forth from the earth in violence, in a manmade or so-called natural disaster, is assimilated, is pulled out and harmlessly released into space by loving the earth. This exercise you did with the Earth Mother a few minutes ago was a small demonstration of how the energy flow is released and dissipated into the atmosphere. Old spent negative energies transmuted into light, and the light coming in constantly from above, so that the planet itself and all parts thereof are transformed. Yet, as you yourself have noticed, there is still so much work to do.

Lightwork reduces the likelihood of a negative force release in some other form. Think how the world peace movement has manifested world peace. And now we decided not to have that war after all. Yes, Nostradamus and many others foresaw a great conflict, a great war in this timeframe. And we decided not to do it. We changed our minds.

Diane: Because the awareness is changing?

Edgar: Yes. And light was brought; it didn't happen. And in the same way, people are deciding they don't need earthquakes. At least not in the same magnitude, not at the same level of disturbance that was originally forecast.

You know what is called plate tectonics? In theory all things are possible, but as a practical level of operation, we can't move the plates. We can't move the faultlines. What we do is balance at the time the point of negative force release. We put positive energy and push against it like shaped plastic explosives, directional explosion, directional charge, And at the point of negative force release, we push, and push the force away to where it will do the least harm.

Diane: And yet people *are* being harmed, people are dying, people are losing everything, everything that they've built in earthquakes.

Edgar: Yes, but look how few people this is passing through as compared to the original projection.

Diane: 25,000 in Armenia?

Edgar: That was small potatoes. We said before that one, as we say now, no matter how bad it gets in the future, remember, it would have been worse.

But let us speak candidly, also. What is the object here? Is it to preserve human life? Not really.

Diane: I think it is to change human awareness.

Edgar: Yes, that is part of it, but ultimately the purpose of this transition, whole transition process, is not to preserve physical bodies. There are too many bodies on the earth. It is to make an orderly transition instead of a disorderly one. Human lives, human bodies, have been recycled for some time now. The preservation of a body is not the object. Is it the saving, the preservation of the soul? No, for the soul cannot be destroyed. The spirit lives forever, and if we lost this planet, what would we do? We would go somewhere else and start over. And it wouldn't even be the first time such a thing has happened in the vastness of the universe.

Is the question whether this will be a planet of darkness or of light? No, that question was resolved in the eras known as World Wars I and II, and in between. It will either be a planet of light or it will be gone. The issue is the physical preservation of the earth planet itself. For in the fourth dimension, the new age, the earth will be physical. It will be a new kind of physical. It will be much like the spirit plane is now. There is nothing available to you on earth planet that is unavailable to us. If we want bodies, we manifest them, complete with all senses: taste, touch and so forth, sight and hearing. If we want homes, they appear. It is creation of personal and interpersonal reality. There is nothing you have that we cannot have. You want sex, you want drugs — if you really want it, you can get it on the fourth dimension. Let us be honest about this.

Diane: Is there a difference in suffering?

Edgar: The suffering is here, yes. There are fourth dimensional planes of suffering also, but this earth will not be one of them. And no one will go there except by their own choice. But in addition to everything available to you, we also have certain nonlimitations. Communication, telepathic communication is universal. Telekinetic travel and transport, also. We have only to think where we wish to be and there we are. This you are already learning to do, you and many.

Diane: I'm quite out of step with the rest of the world, too.

Edgar: You are in step with the future, and this is a wonderful thing. You are a bridge between the third and fourth dimension, and this is a wonderful thing.

What is third-dimensional reality? It is consensus reality. You have heard of "We create our own reality"? Yes, personal reality is interconnected with that of others. The consensus reality for this planet has been redefined. It has already been chosen and is in the process of being manifested. It is called fourth-dimensional vibrational frequency. The new earth will be physical, and the bodies on it will be physical. New kind of physical earth, new kind of physical bodies. You are in the process of adjusting the body as well as the planet for this new physicality, this new consensus reality, which we call fourth-dimensional frequency.

Yet, after the final shift, the projection that there will be ten percent as many bodies upon the planet as there are now is still the same. The question is not whether the transition will occur — it will occur. The question is whether it will be an orderly or disorderly transition, and what will happen to what we call the out-transfers. Those who are moving to new neighborhoods of the universe move for many different reasons. Some because they cannot get their karmic act together in time to graduate to a new dimension. They move to another third-dimensional planet where they will have the opportunity to learn and grow, each at his or her own pace. That is the choice, and it can even be a high choice to do this.

Some will move to other fourth-dimensional planets, positive fourth-dimensional planes much like the new earth will be. They're moving to the same kind of neighborhood, but a different street address in the universe. And of course they don't go at random like seeds scattered to the wind. They go in groups of personal affinity — soulmate groups, oversoul selves, separate selves — however you want to put it. What it amounts to is if we are all going somewhere, we don't go stragglers one-by-one, we hire a bus.

And some will go *home,* after a long time of having helped those who most need help. They are the workers, the troubleshooters of the universe, who will go home, have a nice visit and rest, and most likely thereafter say — "Well this has been wonderful. Now what's my next assignment?" For surely you don't think, in all the vastness of the universe, that this is the only place there is spiritual sentient physical life or that this is the only place that ever got itself in deep trouble.

Diane: I've never felt that way, but the people around me mostly do.

Edgar: Yes, their focus is a little narrow.

Diane: In the book as I've been writing it so far, things are coming together in new ways for me. And I'm looking at this and saying — oh my gosh, is anyone going to publish this, it's so far out? There's a tremendous amount there about other planets.

Edgar: You must reach as far as you can. If you don't the book will be outdated before it goes into print. It's coming so fast, to others as well as to you. You don't want to be behind.

Diane: Whatever's coming is there, but I just need to feel that I am being accurate and clear, and that I'm not making up a fantasy.

Edgar: Oh. The curious thing is how we make up these same kind of fantasies. The channel has often said, if it's delusion, it's group delusion. And if we're crazy, the interesting thing is that we're all the same kind of crazy. The planetary information that you have — we have not read your manuscript or your book — yet, see how this planetary information matches or does not match with you.

We vision, we remember, there were twelve source planets which made physical colonization upon this one, by thought form manifestation and then moving into reincarnation cycle. In addition to these twelve that made physical colonization, there were many, many thousand that have contributed energy but did not make physical colonization. There are ways in which the energy comes in without needing the physical vehicle.

Just as the United States of America is a melting pot of many cultural traditions, so the earth planet is a melting pot of the universe, of many traditions from many star systems. It is very much a mixed bag of energies. You can delineate and define some of the source planets endlessly, for there are an endless number. But the twelve are the ones we focus on, the foundational ones.

The first is Yra-AA I, called the planet of Crystal Light. The Yra-AA system is stellar-political, not all in the Pleiades but headquartered in the Pleiades. Yra-AA I is Pleiadean, as is Yra-AA II, which is called the planet of Amethyst Night. And there is a low gravity planet; the Be-ings are very slender, fine-boned, and they float about easily. They also wear power packs with cross straps across the chest and have two wings on each side that look like dragonfly wings, with electric tracery on them.

Diane: As in the statue of Inanna at Sumer.

Edgar: And they float. Yra-AA III is also called Yrazan, the Fire-Cloud Planet. Rolling clouds of red and orange, many reds — you don't even hav e words for all the colors. And there are little energy units which are the Be-ings, flashing kinetic energy units. There are some who remember being fire spirits in their memories. They were from Yrazan, the Fire-Cloud Planet, before they became incarnate.

Yra-AA IV we call by another name, Tchuith. It is not in the Pleiades; it is called Sirius. O-Sirius and E-Sirius, Isis and Osiris, are from Tchuith. We call them Ah-Tchuith and E-Tchuith. And this is the Planet of High Deserts. The streams are sunken but they are many. It is not a place of privation, as one often thinks of here when one says desert. It is a beautiful planet. And this was the Egyptian foundation source.

Diane: The Dogon call the planet Emme ya, the Goddess Yemaya.

Edgar: Then there is Lyra, which is Yra-AA V, a planet of giant Be-ings much like the physical earthplane persons. The earthplane body for third-dimensional reality was largely patterned on Lyra, so that if the Lyrans come, they will look just like everyone else. And indeed, they are already here. It is on the other side of the Pleiades; it is Pleiades but on the back side, quite far away.

Then there is Mondu I, which is the Blue-Green Water Planet, which like the earth is a third-dimensional vibrational frequency planet of mixed energies. Some of the conflicts which came to the earth were not created here, but came in with immigrants of third-dimensional vibration from other third-dimensional planets. Like the earth, Mondu I has high positive energy, high negative energy, and very much in between.

There is Mondu II, which is the Storm Cloud Planet. Flashing electrical storms all over the surface of the planet — and the people go about on the surface — but their cities are just inside the surface.

There is Taoithan I, which is the Planet of Golden Sands. And this was the second colony placed upon the earth, Taoi, in what is now the Gobi Desert. There are still sonic sentinels around under the sand, and we think these may be coming to light relatively soon.

And there is Taoithan II, which is the Planet of Golden Seas, and is the origin of the Nordic race. Taoithan I is the origin of the Oriental race. You see, each star energy intermixed eventually with earth energy, for there was, oh, the creatures were prepared to receive the implanting of the spirit. And after many relations, the last to be implanted with the spirit was the prehuman earthplane creature, prepared for such implantation by raising the vibrational frequency. Then the spirit was implanted. In Judeo-Christian tradition, this is allegorically expressed as, "God formed them from the dry land with his own hands, and breathed the breath of life into them." Breathing the breath of life, the spirit, into

197

them — that is what we call the implantation of the spirit. This was done by Yra-AA I colonists, Atzlan (Atlantis) being the first colony.

Diane: Not Mu?

Edgar: That was near-simultaneous. Then after a time of further raising the vibrational frequency of the pre-human and now newly-human creatures over a long period of time, this one now imbued with imagination, free will, choice and so-forth, the newly human creature became suitable for interbreeding with star energies. Everyone incarnate on the planet at this time in physical body is the child of Father Sky and Mother Earth. There are sky energies, star energies and earth energies in you and in everyone. Often very mixed sky energies from many source planets. The different ones who tell you "We were the first" — of course they were in their tradition. But perhaps they did not have knowledge of another colony which had been implanted elsewhere.

There were twelve, Taoithan II, the Planet of Golden Seas. Then there is one we call Srignash, which was in the Indian Ocean area, and it is the source planet for the Hindu deities and the many-armed one in particular (Kali). Although it is not Yra-AA Pleiadean nominal, it is very close, it is a stellar political ally, very close kinship with Yra-AA energies.

Then there is one called Systabel, the Three-Pointed Star, which was only about a thousand years ago, and it was an unsuccessful colonization attempt on what is called Easter Island. There were already people there and the newcomers were driven off.

Then there is one other and that is Phra-Shinxth, which is called the Planet of Pain. Much of the violence of the earth planet originated either on Mondu I or Phra-Shinxth.

Diane: Why were they allowed to come here?

Edgar: Because this was designated as a third-dimensional or experiental planet. It was a school where people could come and learn, make better choices than they did last time. It was an opportunity. This planet was, from its beginning, a learning ground. There were high entities, positive entities, and there were some not so high, and there were many in between who came here.

The Ra energies complex — Ra, Brahma, Rama, Yahweh, Allah, Ja, Michael, Raphael, Gabriel, Azreal and certain others — the Ra energies complex was and is the group, the energy-bearing supervisory responsible for this planet.

Diane: What do we call those in female?

Edgar: Technically, all these are male and female. However, the progenitor — if you say the Goddess, it is the same thing — Gaea, Ouiramaya, Isis (who was equal to or stronger than Oriris) — these are female manifestations of that generic non-gender energy. Although Allah has been pictured as a masculine god, Allah is as much female as male.

Diane: Yes, I assume most of them to be like that, but it's not perceived that way.

Planets of Origin

List of source planets colonizations of the earth, by stellarpolitical affiliation:

> (Note: Stellarpolitical affiliations are composed of planetary associations; member planets are often located in far-flung sectors of the Universe and are not necessarily solar or galactic neighbors.

YRA

1. YRA-AA, The Planet of Crystal Light (pronounce E-Ray, E-Ray I)
2. YRA-LA, The Planet of Amethyst Night (E-Ray II)
3. YRAZAN, The Firecloud Planet (Era-jon, E-Ray III)
4. TCHUITH (Sirius Prime), The Planet of High Deserts (Chew-ee)
5. LYRA, The Planet of Giants (pronounce Lie-Ra)

MONDU

1. MONDU PRIME, The Blue-Green Water Planet (Mondu I)
2. SERENIAL, The Storm Planet (Mondu II)
3. PHRA-SHINXTH, The Warrior's or Bloodthirsty Planet, sometimes called the Planet of Pain (pronounce Fra-Shing)

TAOITHAN

1. TAOITHAN PRIME, The Planet of Golden Sands (Tau-I-Than I)
2. VALHUL, The Planet of Golden Seas (Tau-I-Than II)
(Note: A third major colonization by Taoithan-affiliated planets was also instituted by spiritual influx rather than physical landing.)

SRIGNASH

1. SRIGNASH PRIME, also known as BRAMAIHNGNASH. Although Srignash is technically a separate stellarpolitical organization, it holds close ties to the YRA system and could loosely be considered part of the YRA group (pronounce Shri-yash).

SYSTABEL

1. SYSTABEL PRIME, The Three-Pointed Star (solar system with only one habitable planet), near the black line (Sista-bel)

NOTE ON NAMES: These source planet names can be transliterated or translated (semi-idiomatically) in a number of ways and have historically been called by many names, some of them similar and some quite dissimilar.

FURTHER NOTE: The sources for spiritual colonization have not been mentioned in this list (except for noting Taoithan III). There are a great many of these.

Edgar: That is correct. It is the perception that needs correcting. But the fact beneath the perception is not gender specific. Yes, it's androgynous. There are many planets where there is no gender differentiation. It is all androgynous, all one. There are also planets where there are three genders, and it takes three in union to reproduce. But we point out that this gender differentiation, while it is not unique to the earth planet alone as some people perceive it to be, is also not a universal concept. It is neither unique to the earth nor universal concept.

Diane: We have enough problems here for those of us who prefer our own sex.

Edgar: You feel this way because you have chosen to work in the area of female empowerment. This is an excellent choice which we affirm and confirm.

In speaking as Edgar Cayce, this is an appropriate time for me to acknowledge the feminine aspect and history that I share with other aspects of myself, my greater self. There is no one among us in the Circle or in the light who is not female as well as male. It is not, how shall we say, that a lack of experience from both gender viewpoints, would hamper someone from doing lightwork at the higher level.

Diane: Seems to me that the great learning or the great tragedy of earth or our time is that earth has to oppress one over the other. And it's usually, at this point, been women who've been so terribly oppressed.

Edgar: But that's what you're in the process, we're in the process, of overcoming. Yes, there have been times when it was the males who were second-class citizens.

Diane: Were they ever treated the way women are — comparably? Were they ever treated as bad comparably as women are?

Edgar: No, not on this planet. It has been done elsewhere. But let us be honest — on this planet the males were not treated that badly when God was woman. That was one of the consensus, one of the parameters of consensus reality which was chosen for this planet as a learning experience. And those who gravitated toward this place had need of that experience.

Diane: Then what I call the matriarchal reality, which does not mean women dominating men but means a balance, is the balance that is desirable.

Edgar: Yes, it is the balance. Where there has been such male domination for so long, obviously there must be a balance problem.

Let me tell you one insight into one way in which this was manifested on this planet — the male domination. In the Peruvian-Atlantean connection, the Be-ings were Lyra and Yra-AA origin. They were physically narrow-shouldered, narrow hips, long arms and legs, very long slender fingers, feet and faces. And they were androgynous. They are and were beautiful, and they came to this planet. And when the interbreeding began, because of the narrowness of the hips of this androgynous Be-ing, when they interbred with human energy, it was male star energy to female human energy, because the star Be-ing could not safely bear the earth child.

After many centuries, we looked around and realized that because of this physical accommodation to body structure, whereby it was predominantly male sky energy to female earth energy, Father Sky and Mother Earth, that somehow the spirituality aspect was not as strongly implanted in the females as in the males. And they had more sky energy in the males, they were dominant. Not cruelty, simple dominance. And at that time (and there is some painful memory here for you) we recognized that in order to bring spiritual reality at comparable strength to the female line, some would have to go directly into the earth bodies. And there were those who chose to do this, to bring the spirit to the female line, the stronger line. They entered into the female earthplane body and there is memory in you, my daughter, of what it is like to be a totally spiritual Be-ing in an animal world.

Diane: I still feel that way at times.

Edgar: This is the origin of that feeling. You were one of those that chose to do this thing.

And the first of these was called Ouiramaya, remembered as Maya, the Great Mother. And it was through these re-spiritualized females that female spirituality was restored in comparable degree to (that of) the male. Yet there had been time lapse, and there was much work to overcome male dominance. This was particularly true in some parts of the world, not all. Particularly true in American Indian culture, in parts of the Middle East, and in the Orient; not as true in Egypt and some other parts of the Middle East where the Goddess energy, the strong female spiritual presence, had come through star people not configured like those from Yra-AA and Lyra with narrow hips. The configuration of the physical body assumed by star people was decided in part by what they looked like at home.

Diane: Would it not have been simpler to rearrange the physical?

Edgar: Very likely so. But remember, we said this was an experiential planet, which also means experimental. Oh, hindsight is a wonderful thing! If we had it all to do over, there are any number of corrections that could be made. Yet, ask yourself how many of these corrections would also involve impinging on someone's free will. For free will is not only the simple choice between right and wrong, and do this and don't do that, it also implies imagination and self-expression.

Diane: What were the creatures that the early creation stories talk about as being half-human and half-animal? Were they truly animal-shaped bodies?

Edgar: They were several things that have been remembered in this form. One is that they were experiments that were discontinued. Oh, another thing is that they were replicas, attempted replications of some remembered forms from other planets. And the third is that they are representations of the energies of those whose spirits moved from human to animal form easily.

Diane: Why were they abused so terribly? Or am I mistaken in feeling that that is what happened?

Edgar: They were abused at the latter part of their existence. This was a very short historical era because these person-creatures had much control over their own existence, and they didn't have to stay and be abused.

Diane: And they finally chose to leave?

Edgar: They chose to disintegrate that body form and manifest the spirit in other ways.

Diane: They did exist, then.

Edgar: Yes, briefly.

Diane: I have an image of a changeover from — I call it a matriarchy, you may not use that word — a spirituality-centered way of life to what we have now. My understanding of it could be wrong, but please correct me. Did that involve the development of Atlantis at the expense of Mu? Or at the end of Mu?

Edgar: You have a very common misperception. It is not without some foundation, but we consider it a misperception. Atlantean culture was hi-tech and the Lemurian was low-tech. Lemuria was only in part female-centered. That was late Lemuria. Early Lemuria was not even androgynous, it was asexual — it was spiritual reproduction rather than physical bodies. But that was very early Lemuria. Atlantean civilization and Lemurian civilization: Lemuria...Oh this next shift coming up is the fourth. Lemuria went down in the first and second. Atlantis survived the first and second and went down in the third.

Diane: Okay. That came through for me, also.

Edgar: By the time of the third, Lemuria was only remnants and blended with Atlantean.

Hi-tech Atlantis — what happened? We knew the shift was coming and we told people, "Go to high ground." We knew where it would be, what was happening, and we said, "move now." And everyone said, "Oh, we'll wait to the last minute and then we get in our little air cars and air boats and we go for higher ground." But the transition, the shift, is geomagnetic in character — realignment of the polarity of each molecule. And approaching the shift, the geomagnetic grid was disturbed. And the air cars and air boats and all the full technology of Atlantis was geomagnetic based, controlled by crystals. So it didn't work. When the power fails, who do you blame? The electric company. We didn't do it; we just got blamed!

Now, Lemuria, which is low-tech. Instead of saying "we jump in our cars, we jump in our air boats and go for higher ground," they said, "we will teleport. No problem. We know how to do this thing. We will teleport." What kind of energy is teleportation based on?

Diane: Also geomagnetic?

Edgar: Yes! So that also didn't work. And it was too late to walk.

Diane: But they knew, they knew when it was coming?

Edgar: They knew it was coming. And some of them had sense enough to start walking early on. Or to teleport themselves before the geomagnetic grid was disturbed. But the essential reason why Lemuria went down was the same reason why Atlantis went down.

Diane: The culture?

Edgar: It was the shift, the polar shift, the realignment of each molecule, the electromagnetic force. It's what is happening to the earth again. And this technology will fail. But here again we say, it is not a question of whether the transition will come, it is a question of whether it will be orderly or disorderly. And new technologies are already being developed to fill the gap.

Diane: Can you talk more about what's coming?

Edgar: Yes. That event some people have called the Second Coming, the Christ consciousness manifesting, was originally projected for 1998, to be followed by the shift of the axis in 2001. The Second Coming manifested ten years ahead of schedule, in 1988. And let us say here that we're not talking about Christ as a male, we're talking about an energy called the Christ consciousness which is as much Earth Mother as it is any other energy.

Diane: What does that mean when you say that that came to this planet?

Edgar: It started in 1988, ten years ahead of schedule. The first stage of the project being the rising light consciousness in the hearts and minds of the people.

Diane: The Harmonic Convergence.

Edgar: Yes! So, it is ten years ahead of schedule. That does not mean that the final shift comes in 1991. It means we buy time to do more work to buy more time. And the shift, instead of being in 2001 will perhaps be as late as 2020. Or perhaps as early as 2012. For we have been so successful in raising consciousness — all of us together, yourself of course very much included — that the healing is ahead of schedule, well underway. And it may be that before 2020 we say, "Well, it's not perfect but it's as good as we're going to get it; we might just as well finish up now." But our best guess at this time is it will be 2012.

There will be certain preliminary events before then, some of them more dramatic than others. Much has to be defined by consensus reality. And then there will be, at the final shift, in a single twenty-four hour period, a tipping and shifting of the axis. It is not actually the core of the earth that will move in relationship to the universe. The core will be where it is now. But the crust, there will be big bubbles in the core that comes and breaks and jolts loose the crust, and the crust will float around the core. And then reattach. And that is what people call the shift. And afterwards the magnetic flow that is now in the Northern Hemisphere will be in the southern, and vice versa.

Diane: Will there be that much of a shift?

Edgar: Yes, it will be a complete reversal. This will be the fourth time this has happened. Ask the geologists. They'll tell you this has happened before.

Diane: And it will also accompany great destruction?

Edgar: Yes, that is true. But the preliminaries will also be great destruction in one form or another. We foresee a shift of focus from earthquakes (although there will be more), to manmade disasters — oil spills, chemical contamination, terrorism, random violence. All of these are already on the upswing, and we foresaw this sometime ago.

The contaminants must be purified somehow. They must be folded back under the crust so they'll be broken down to atomic structure level where they'll do no harm. And at the same time, new resources come up from the crust, up through the crust. There should be already beginning, rising of the earth under portions of the Atlantic near the North American coast now, in the north Atlantic near Iceland, Greenland, and such. In the Pacific — yes, they should be there. We haven't read the papers.

Diane: A new island has come up near Iceland.

Edgar: We have been speaking of this for some time. We are glad to hear that it has been reported.

Diane: They named it Surtsey.

Edgar: Oh yes, we know about that one, but there are more. Many more, much stronger. There would be also a new rising near what was once Krakatoa, and there will be a sizeable land mass there by the final shift and shortly thereafter.

Diane: That's in the South Pacific.

Edgar: Yes. There will be a new bay in the Gulf of Mexico, where portions of South Texas, Louisiana, Alabama, Mississippi and Arkansas are now. Memphis is still expected to be a seaport. That is update, but it correlates with the projections that were given some time ago.

Diane: So those cities will be destroyed?

Edgar: Many, yes. But there will be warning. People can readjust. There will be a rising in the Gulf of Mexico, and right now California looks better than it has in a long time. We think California might be preserved. Not surely, but a good chance of preserving that part of the plate. Alaska is looking worse. The oil spill caused priority to fall for preserving resources in Alaska, resources being trees and natural resources. Therefore, some of the negative energy that would have focused on California is shifting to Alaska instead. And we expect seismic activity there relatively soon.

Japan is still in danger, but is looking better than it has in a long time. You know how the crust, the plates are such that one folds under another? Japan was expected to be folded not only under the water, but under the Asian land mass. The only way to prevent this from happening, we have been in the process of working on for some time. It involves controlled seismic activity. We take Japan and we separate it from attachment to the crust, float it on the liquid just under the crust farther away and reattach it at a safer place.

Every time this controlled seismic activity is triggered, there is the very distinct danger Japan will all go down. It is a calculated risk. If we don't do it, it goes down for sure. And if we *do* do it, *maybe* it is preserved.

Diane: What's the potential for preventing destruction — loss of life, major damage — by the lightworkers that are working toward changing the things that are wrong?

Edgar: Excellent potential. We go you one better. In the past, how have you moved from third dimension to fourth? You died, you left the body behind.

Diane: I don't! I don't know how other people do it.

Edgar: Ah ha, you make a good point. Let us re-express it. What has been the traditional method for others has been death and leaving this planet.

At this time, it is not necessary to die to go from third to fourth dimension. You are already doing this, as many are, and you are adjusting your body so that the body can move directly into fourth dimension without experiencing a death crossing transition, without being left behind.

Diane: The body will go too?

Edgar: Yes. And will still be able to function in third dimension as well as fourth. This is what you are doing now. You are swinging back and forth. It is not from low third to high fourth, they are close together. And you are moving from high third to low fourth. There is much interchange between, and lightworkers, many people, everyone who chooses, need not die, but can take the body directly into fourth-dimensional frequency. This is why you have fitness craze; they are preparing the body. This is why you are releasing old blockages; you are clearing the way. And in this high third/low fourth, the earth has come into alignment with the lower astral planets. And this is why disturbances, spirit attachments, and all sorts of old garbage is manifesting, for the purpose of being healed.

Now, the lightworkers at such time that the shift will come — we're moving from third dimensional to fourth, finally the whole planet, yes? For the whole planet is already halfway into fourth, or better than halfway — the lightworkers will already be in fourth-dimensional vibrational frequency. And they won't be troubled by that which you perceive as disaster. They will be helping other people who still need to make that transition into fourth.

Diane: That's what a lot of us do at night and some of us aren't even aware we're doing it.

Edgar: We're all in this together. We ride the storms, we adjust the energies, we commune and communicate with each other. Very many are engaged in this.

Let us talk about some practical realities. Think how useful it would be when the lava flow is coming down your street if you can levitate. You raise yourself above it and it flows by underneath you, yes? But you need not wait for the lava flow to come down your street to find levitation useful. Let us assume there is some very physical, mundane, earthplane person chasing you, intending to do you physical, mundane, earthplane harm and you can levitate. But you need not raise yourself all the way above him so he runs by underneath — you get this far off the ground and he becomes alarmed and will run away.

Many of those tricks — mystical, magical, theatrical tricks — are practiced for the future. Levitation, dematerialization and rematerialization: we're talking about fully functional ascension and interaction between dimensions.

Diane: The body goes too; it's not only the soul that's going?

Edgar: The body goes, too. It is not only the spirit, it is the body also. You are manifesting fourth-dimensional vibrational frequency by your spirituality.

Diane: And when I teach others to do something like distance healing, I'm also teaching them this dimensional change, too?

Edgar: The healing of the body is preparation for it, to become suitable to flow into the other dimension. Let us not overlook the fact that those who choose to discard the body may be making a high choice. Don't make them feel like second-class citizens. Healing is not always the preservation of the body.

Diane: Healing to me involves balancing the spirit, too.

Edgar: It can involve that also. But don't focus on body alone. Healing of the spirit — too many people say healing and they think only the body. There is healing in which the body is still discarded. If someone has chosen to make the death crossing transition, for whatever reason, and that is true to their choice, your lightwork won't stop them. It will make them more comfortable in the process, and this is a wonderful thing.

Diane: My concern with so much of what you talk about in earth disasters, mass deaths, in an earth change setting, is not the deaths themselves. It's the suffering and it's what the people left behind have to deal with.

Edgar: Exactly. There is no pain in death. The pain is in preparation. Where there is enlightened understanding, this pain is greatly lessened. There is a certain sadness of separation, but we often bring the assurance, the reminder, that the separation is neither permanent nor complete.

Diane: In this culture, that's an awareness I have, that's an awareness a few people have. But only a few.

Edgar: And you are in the process of sharing it; that is part of enlightenment.

Diane: What's this New Age that we're going into? You talked about a Christ consciousness, an awareness of the spiritual.

Edgar: It is simply a name that many use to describe spiritual awareness of a higher level.

Diane: And Christians would call that the Second Coming?

Edgar: Yes. There will be these physical manifestations. They will *see* Christ, and there will be the Buddha, the Mother, there will be many.

Diane: And along with this there is also a deterioration of the quality of life on this planet — more violence, more pollution?

Edgar: Yes, this is the stress manifestation of rising vibrational frequency which, no matter what path you are on, is an acceleration. The violence is increasing but these energies, when they go out, they don't come back. It's not the same way. They are being withdrawn from the system.

Diane: But they're still manifesting.

Edgar: Yes, because we are still in that stage. But they are being withdrawn. They are being shipped out. This is the out-transfer we spoke of, of moving to other neighborhoods. And that everyone is striving to be eligible to move to a good neighborhood. Everyone goes to the personal reality they have created for themselves, and not all of these are pleasant. But they are being shipped out.

Diane: And then the next thing is the shifting of the planet and the destruction that goes with that?

Edgar: The shifting of the planet: it is either an orderly shifting or a disorderly shifting. Where it is disorderly the term destruction is quite apt. Where it is orderly, then there is less that you would perceive as destruction. It is simply renewal, realignment, readjustment. If everyone will listen and plan and participate in the orderly transition, then there will be no disorder, there will be no destruction. But not everybody listens at the same rate, at the same pace.

Diane: Does it have to be everybody, or enough?

Edgar: It is called critical mass. It has to be enough. And then the others have to have their opportunity to move.

Diane: And then after the pole shift, then there's the rebuilding and a new way of doing things.

Edgar: Yes. New technology and the full new dimension — which will be too tame for lots of folks like you. You'll want to go home and see what next assignment comes to you.

Diane: Am I going to die in the pole shift?

Edgar: Die? You aren't ever going to die.

Diane: In the physical.

Edgar: Not unless you choose to. And we think it highly unlikely you would manifest that choice.

Diane: I feel so trapped in a body, but suspect that a lot of my longing to get out of here is not to die, but to move into that fourth dimension. I never saw it that way before.

Edgar: Leaving it behind is not the same thing as dying.

Diane: That's what I'm starting to understand.

Edgar: It's called "body optional." It's just like it is for us, bodies are optional. And you will manifest that reality where the body will be yours for as long or as much as you need it for. If you set it aside for a time it won't decay; you'll still be able to put it back on.

Diane: I have no fear of death, I never have.

Edgar: Nor should you. There is no pain in death. There can be pain leading up to death, but there is no pain in death.

We'll say goodbye for now, hoping very much to have more opportunities to speak of these things again. For it is our pleasure, truly to speak with you.

Diane: Thank you.

The Channeling Sessions

Tanith — June 21, 1990

Diane: What condition will people and the earth be in and how will that change? What will earth culture, human culture, evolve to because of the earth changes?

Tanith: You know the changes the earth is going to go through, and what you're looking for now is what society's going to be like and what will we be left with?

Diane: Right, how's it going to be different from what it is now, how's it going to be improved, is anything going to be worse, is everything going to be better?

Tanith: No, everything won't be better, and that's going to be by definition. How do you define better? There's going to be some people who are going to be very very content with it, and there'll be some people who are simply going to bemoan all that they've lost. And there'll almost be like a second cleaning. This period now is a period of deciding who are the ones who will move forward and who are the ones who won't.

Diane: That's what's going on right now.

Tanith: But there'll be a second one because many of the ones who feel that they can do it — feel they can move forward, and that they are ready for it and capable for it, and would be happy with it — will get there and find that they were wrong and they can't. And all they'll do is miss the things they had. They didn't realize that changes also meant that maybe the VCR doesn't work, or that they didn't realize some of the things that will be gone.

Diane: What things would be gone?

Tanith: A lot of the things that are real luxuries that are considered essential. And I don't think they'll be totally gone. I don't see that we won't have electricity, I don't see it the dark ages, but I see that conservation is so incredibly important that we don't simply waste it. We don't waste the electricity, and there'll be other things that we'll be doing so that, uh...the movie industry won't be as big. We won't be making the movies so we won't have a new slew of movies every summer and movie houses to go to. There still might be movie houses but it's not going to be the brand new movie every single week coming out hi-tech, because the tech will be going for something else. Some things in the home — home will be pretty much okay — it's not the dark ages. It's not like we're going to lose washing machines and dryers, but there's going to be less use of like VCRs cause there'll be less tapes to play in your VCR, and the TV won't be the main source of entertainment anymore.

Diane: Sounds like an improvement to me.

Tanith: Yeah, but for some it won't be. For some it'll be very saddening. For some it'll be a step back. There's something about creativity. The children today have lost creativity, there's no creativity, they're losing it. It's being trained out.

Diane: Right, right.

Tanith: That is something that needs to come back, for survival of the unity of this planet. Humans can leave and the planet would survive. If humans want to stay and have the entire planet and planet system survive, then some things about humans have to change. And one thing that has to change is the creativity. It needs to be bred back into humans. To do that is to take away the things that deprive and rob that creativity.

Diane: Creativity's punished now, will it be then?

Tanith: No, because that will be survival in many ways. Storytellers might be more the entertainment than TV, and books more so than TV, and artwork. A lot of the creative kinds of things are the things that will be more appreciated, as opposed to the technological. And if it's technological, it's because it's creative technology. It's technology that fixes things; it really *does* something.

Diane: Like?

Tanith: I don't see us being stuck without any kind of motor transportation, for example, but I see the transportation changing into one that runs on something other than fossil fuels. So somebody who's creative invents a solar car, and someone else invents a wind car, and somebody who runs their car on apple juice or something is going to be very much appreciated. That kind of technology, then, is healthy for the planet and allows humans to have some of what they need, which is that freedom, that movement. And they need that still.

Diane: So is it that these things will be gone? Or is it that people will be aware of the ecology?

Tanith: Well, some things will be gone because they won't be useful. Gas is going to be very very expensive, gasoline. And so you might still own your car and there might be gas stations, but you might have to be a millionaire to drive it, because it won't be accessible like it's accessible now.

Diane: Will there be any transportation that's accessible?

Tanith: Yeah, I think that there's going to be solar cars and wind cars, and I don't know about planes — something about planes, there's something different about planes. Almost like we have this leap in technology that occurs in like ten years or twenty years. This leap will take us into the Star Trek generation practically, this leap will take us to that different form of energy that allows us to commute in a form that's something else and non...

Diane: Antigravity?

Tanith: Yeah, yeah.

Diane: Others have been saying that, too.

Tanith: Even some of the car vehicles will be doing that, too. Just regular vehicles, motorcycles, things like that will be not with wheels, they'll be antigravity. They'll be skimmers. We'll just take this leap in technology, and that kind of creativity will be really really appreciated, will be really valued.

Diane: The difference in how children will be raised, what will that be like?

Tanith: It's going to be smaller groups, and I think what we're going to find is that the schools will be run by parents. There will be schools, because schools are just so efficient, and because schools allow so many other adults freedom. But the teachers will be parents of the students, and what we will do is take the people that we highly value. Older people, too. We're going to take old people and make them teachers, like when they retire and don't do other stuff anymore. We make them teachers because they're good for the kids and safe with the kids, and because they have so much to share.

Diane: But now the kids are not safe for older people. Kids are violent now. That won't be the case anymore?

Tanith: Well, that will be a different batch because violence can't exist in that world. There's going to be segregation.[1] That's what I mean when I say there's going to be a second weeding out. We're going to weed out a lot of the people who can't move forward. There's going to be a second weeding, and that second clearing is the ones who thought they could do it and now get there and realize they want the old life. They're the ones who still perpetuate the violence and the crime and the...

Diane: How will that second clearing out happen?

Tanith: Almost like the societies segregate themselves, that you have neighborhoods that are of "those" people because it's almost like a racial issue. Now it won't be racist, but it'll be like that. But the ones that like the violence will sort of segregate together and the others won't. And the ones who are not with them will be the ones who survive, the ones whose neighborhoods can survive, the ones where dis-ease does not — it's almost like the other ones forget. They just go all the way back; they slide all the way back. They forget all the things that kept them healthy in living through the times and then they slide all the way back. See and it isn't violent destruction, it isn't like the world is going to stop. It isn't going to stop. There'll just be less humans.

Diane: How much less? Some people are saying only 10% of the current population, world population.

Tanith: I don't know, I don't really feel that little. I feel sort of like 50%, because the world could handle that. The world could handle about half. But in some areas — because it's not going to be 50% spread out evenly. So in some areas it's going to be pretty pretty massive, and it isn't going to seem fair. I guess that's the other thing, it isn't that we're wiping out any one culture, but you might find a lot of death around some cultures. And the reason is because they have destroyed more in less time than even this country. They've been pressured to develop, and there's less left for them to be able to save, and they're too far away from understanding. They're still moving in that direction of destruction. And so to save that area it might be necessary that more of one culture die. And it isn't any kind of moral judgment, that they were better or worse, but that the society as a whole was more destructive. So there might be some excellent people in there. And it has nothing to do with color or race or religion or creed, but if that group as a whole is more destructive.

Diane: But there will be racial equality, finally?

Tanith: Well, you know, I'm not sure. People are people. Black horses don't like white horses. I think what you'll find is a greater equality in the numbers of races worldwide, but possibly not the mix. It may not be totally mixed per country. And it may seem not fair, and again it doesn't mean that the people dying were bad people, or that the race is a bad race but that people are still going to remain people in many ways. They may be more tolerant, and that will continue to grow. Equality is a weird word because we're not going to see world equality right away, where everyone has the exact same standard of living. Not right away.

Diane: But will we stop seeing the kind of thing where we're putting you down just because you're Black, or you can't have this job because you're Black, or you're not good enough just because you're whatever color?

Tanith: No, that'll be pretty past. It will be one of those things where if it happens it will be recognized fairly quickly and the community as a whole will not allow it.

Diane: Good.

Tanith: And the person who is like that will find that they are ostracized. They will have the pressure on them to match the whole, and the whole will be one that is a lot more open.

Diane: How about women's status?

Tanith: Women's status is going to change considerably, 'cause part of what's happening is a lot of the men are going to die. And we're really seeing that, that a lot of men are dying.

Diane: More than women?

Tanith: Yeah, more than women. And there's going to be some countries where it's women, children and men, or maybe more women and children, but overall we're going to see a lot of deaths of men. We're also going to see a lot of women who just stop listening to men, who just say, I'm not going to listen to this, I'm just going to feed my children. I'm not going to deal with this, I'm just going to take care of what I have to take care of. I'm just going to farm my little plot of land and I'm not going to deal with this. So we're going to see a change in that women just stop listening, stop allowing it, 'cause women have participated, they have allowed this. And they will just simply stop allowing it. It isn't violent. It isn't a violent change, it's just, I don't have to do that. And no you can't have my money to put into your bombs, I'm not interested, forget it. And by that point there won't be any way that the government can enforce the collecting of taxes because they'll be breaking down.

I think the American culture is going to allow the country to continue to run, but under a much less powerful status. I think that what we will do is allow there to be a Congress, and allow there to be a Senate and allow there to be a President, because it worked so well for so many years and it makes us feel good to know we have that. And we just listen a lot less and we give them a lot less power, and they're more a figurehead like the queen in England. It's more like a figurehead, more like an honor of letting them go and sort of play at doing that.

Diane: How will the country actually run then?

Tanith: The country's going to have more autonomous states and towns and city-states and there'll be basically that cover over it that says, okay we are all Americans and we will all have that in common, and we'd all fight for something if it came to it but it's not going to come to it. And basically it's going to be women saying we need to feed our kids. I think it's really going to come to that, it's going to come to a food issue. We need to feed our kids. We aren't interested in the rest of it, we need to feed our kids. We need to heal our kids, we need to have health care for our kids, we need to take care of our kids, we need to have healthy, smart, educated, kind kids. And that's what we need to do. And much of what the United States government does will just be ignored because it's international, and the international's going to stop a lot because people are going to say, heal here first before you heal out. And a lot of the countries we're giving money to may not be there anymore to give money to, and we won't have any money to give them anyway. So they may make policies about how they talk to other countries but it won't really matter. So at most the government will still come back down to city. Most of the government now that affects people is city-run, state-run, and it's still going to keep doing that falling down.

Diane: Will government have more compassion in how people are living?

Tanith: People will have more compassion about how people are living, and government will just be ignored if it doesn't. People will just do it themselves. There'll be more backyard coalitions that form to plant backyard gardens to feed all the neighbors and make sure that every old lady in the place has a house and has food, and no one will ask the government. It's like, can you get government aid, well no, but why would we need it anyway? That's not what we need, we need neighborhood aid.

Diane: Then people will take care of each other?

Tanith: People are going to start taking care of each other because they'll understand that it benefits each other. A lot of people who haven't had jobs for ages may find that they are incredibly needed in that coalition pulling together. Some of them will feel like we've really declined because now we're back to sort of just living. It's not a survival issue, it won't be like we're all going to die off if we don't do this, but just that everybody's taken a step back and said, well let's enjoy life. So a lot of the intense competition in businesses, in money, in government and technology won't be there because no one'll really care. And in technology we'll take that leap, and that technology will partly help save us, because it will be that technology that says, we can do this another way. Also when we do that technology leap I think there's going to be a lot of people leaving this planet.

Diane: Because of the technology leap?

Tanith: Yeah, because they think it'd be so fine to go to another planet.

Diane: Oh I see, okay!

Tanith: So they leave, and it's interesting because there's a few that already know that no matter what they'll stay, and they're the caretakers. They're the ancient caretakers who've come back in this life to stay. And they're kind of like the old Ents in the (Tolkien) trilogy, and were there as guardians and they're very very ancient. They're incarnated as humans but to stay perhaps what they have to do is simply Be. Their energy of being on the planet at this point of time is like an Ent, is like the energy of being around the forest, the one who walks the forest and knows that as long as he is there the forest will survive. And that's what these people are, they're simply there and by the grace of their being there, by the grace of their energy being on this planet, they are helping feed this planet. They will be the ones who are never tempted to leave.

Diane: How about religion?

Tanith: It's going to get very diffuse and variable; you're going to have many many little religions. There's actually going to be a bad period before the end, where we have a lot of Bible thumpers and soapbox ministers screaming and stuff, the end is coming the end is coming. And they're right but it's their end. (To Onyx the dog: Okay, I'll tell her. I'll tell her. Can you sit down?) They will be some of the ones that go. There won't be leaders of religion to lead religions anymore.

The Vatican is probably going to get hit by some sort of subeterranean earthquake, deep earthquake — and I have no idea where this is coming from because I don't know the fault system around there — but there's something going to happen that will shatter part of the Vatican City. And what will end up happening is some of the secrets of the Vatican which are old Pagan secrets are going to get revealed, and people in power there just won't be in power because some of the corruption will be revealed. It may be like a spiritual earthquake where there's some revelations about what's really going on there. And when this Pope finally dies they will not be able to pick another Pope, because they'll pick one and he'll die. Just because. No one will kill him, it'll just sort of happen.

Diane: This current Pope?

Tanith: Yeah, and there'll be a lot of unrest because of some of the things people are finding out about the center core of these religions. A lot more disclosures like the Tammy Bakker thing, a lot more disclosures about that. Then eventually what we get to is very personalized religion, where there aren't very many great leaders. You'll have people building little community churches or worship places and some of the religions will last. There'll be a form of Christianity that lasts, and there'll be a form of Judaism that lasts, and there'll be a form of Zen — they will last. They will last in a form but they will all start to integrate more of a worship of the earth because the earth has almost destroyed them. In the people's minds this will start to integrate more. Not because of some great spiritual awareness but because the earth has shook them up and scared them enough that this is going to be spooky.

I also think there's going to be what some people will call a Second Coming. It won't be Christ, it'll be a Mary experience and probably in South America, where we have another Mary Magdalene that shows up like we've had through the centuries. And part of her message will be, "I am angry. I am angry because you never really listened. You listened to my son, you didn't listen to me, you twisted it all. I was the one, it was not my son. I told you to love and I let you borrow my son to see if you could do so and you couldn't. I told you to be joyous and you didn't listen." She'll be coming back with an angry message.

Diane: This is like appearances of Mary? Like at Lourdes and Fatima?

Tanith: Yeah. Right, yeah. And probably in South America because there it'll be believed. It's the first place that will be believed, so that's where it'll happen. And the message is going to be that the Mother is angry. The Mother is angry, and you'd better understand the Mother. Probably some Indian women and Black women and other women of color — it won't be a White woman — I mean it might be one, but it won't be all White women in these appearances. It'll be women of color.

Diane: The Earth Mother herself will appear as a woman of color?

Tanith: Yeah, because we have to deal with the Black Isis, the Black Kali, with the dark, secret, powerful side of the Mother. This will be the Crone aspect of the Mother and beyond that the primitive aspect of the Mother that says, I can also kill. And I will do so to save lives. I will do so to save the essence of what this is.

Diane: So this would in essence be a Goddess? A Goddess experience: What will happen to the Goddess religion?

Tanith: The Goddess religion will be fine. It's never going to be mainstream. I think we're going to find an earth religion, sort of an earth kind of religion. People still might call God "he" and that's okay, and the Earth Mother recognizes that. But the Goddess religion will be the permeating force that kind of keeps things moving. It still might be the minority religion but it will be okay because it will be a religion that's more accepted. It'll be just accepted as a minority religion and it'll be something that in many ways helps guide a little bit. And it's okay with the Mother that it's a minority religion because simply those people who believe it and live it and create that essence of that belief system and relive that mythology — that is the energy that's very very important. It may not be important for those people to be mainstream and to spread their message to other people, it's simply important for those people to exist and to help affect the whole by the message they give.

Many of those people will be women. It will finally be women who have come through enough cycles and turmoils and changes to be strong women, and to be secure women and to be who they truly are, and to transmute so much of the pain into joy and so much of the hate into love, and so much of the energy that's around into healing energy. These women transmute it through themselves, can actually be transmuters, so that energy comes in and goes back out in a different form, simply transmuted, and they are not harmed. They will transmute it for the earth so that healing energy is transmuted into the earth, transmuted for others around them.

We might find a real upsurgence of almost like the Delphi prophetesses. Not gurus, more just that awareness that somebody who could be like that is there and sometimes we need to touch that. Those messages may not get spread in the newspapers, but they would get spread via that energy entering the world and then disseminating around the world. It'll be energy transmissions and we'll find that many women — and men, too but women are really the leaders of it — are going to be really incredibly psychic. Not psychic like great tarot readings, but psychic like talking from Pittsburgh to California and knowing, just knowing.

Diane: Telepathy?

Tanith: Yeah. Just knowing that this is the message and here's what we have to do. And letting those messages circle the earth so that as messages of healing and love and ecology and growth circle from woman to woman around the earth, they create an energy field around the earth on the psychic level, and that itself is healing for the earth and affects everybody else. Kind of like when people are under an electric wire something really does happen whether they know it or not. So if they're under a psychic field that's communicating this from woman to woman (and man, too) but all the psychics around the world, and they're all communicating that to each other, it creates that kind of energy field.

And Onyx said something about animals. She said that we were going to understand animals a lot better and that dogs will be much more important.

Diane: Oh, there's nobody more important than dogs. We know that!

Tanith: People will realize that, yes dolphins do have great gifts, in fact dolphins are more intelligent than humans and they are part of our teachers.

Diane: If we had sense to listen.

Tanith: Yes, they're really really patient. They are from another planet, their original core, so they won't die out. They may die out on this planet, but they won't totally die out, and they won't die out anyway. It'll all stop before they all die out. But they are our teachers and they are observers, too. They're so incredibly advanced that what they have is a sense of humor. They're so far beyond us in intelligence, it's like no sweat. For them the message at the end of it all is very simple: have a lot of self-confidence, be who you are, love yourself, live in joy, take care of the earth, and have fun. It's like Doug Adams' *Hitchhiker's Guide to the Galaxy,* you're just supposed to have a good time. The message is just "enjoy life."

Diane: Stop panicking!

Tanith: Yeah, right. There's going to be lots closer bonds between animals and people. Definitely the whales aren't going to die out and definitely the animals aren't, and that's one reason that some of the destructions are coming by water. Because destructions by water don't hurt the water creatures. If the water level rose, even if. See and there really isn't global warming, that's really interesting; there really isn't, it's normal global changes. We are in a warming period and in many ways we caused it, but that warming period is just a natural cycle of the earth. If it really does warm up enough to melt the glaciers, the animals in the sea will still live. The sea will be okay, it'll be refurbished with the fresh water from the glaciers and there'll be a higher sea level, and they'll be protected from the extreme heat. They will survive that. High altitudes will survive; low altitudes will have water and the animals will run to many of the higher altitudes. I don't see everyone being a vegetarian, either. I don't see that happening, but I do see more respect for what you kill.

Diane: Less waste? Less cruelty?

Tanith: Yeah. Less waste. Less cruelty just because the people who are cruel aren't going to be selling their product. In the American Indian tradition, there is a part of the tradition that said there was another time for them, that they are coming to that, but many of them have lost it. And there's great loss in those people.

Diane: They've been damaged by the culture, too.

Tanith: Right. But they have not held true to what they needed to do. Their religion will hold and those of them that can hold onto that religion will be alright, will help with the changes. They also will be more accepting about others coming in, understanding that at this point who was on this continent first is a silly question. Anyone can practice this religion if they live it and believe it. They will be more open.

Diane: What about contact with other planets? Will that happen?

Tanith: About the time we make this leap in technology, a lot of people are going to think it's got to come from somewhere else. There are nice planets out there and there are not-so-nice planets out there, and I feel that for a while we will be lucky because we are being watched. We are in a way being cared for as children of another planet that can protect us. Very very far in the future we may have to deal with not-so-nice planets. But at this point we just don't have that concern.

Diane: Is it going to be the same thing all over again like between nations of the world? The wars and politics and economics and all the negativity we've got now that we're trying to get rid of?

Tanith: I don't think so. I don't think that's going to happen, 'cause I think we'll have learned. The people that survive from here and move onto that will have learned. They aren't going to play the same games. It's always two interacting and you both have to play the same game. And it won't be played the same way. That's very far away and really not a concern now; it's too far away and that future is not totally created yet.

Diane: And that can change?

Tanith: That can change. It's not totally created yet. It's not really a concern and not something to really be thought of yet.

Diane: The near-to-come contact with other planets, what will that be like?

Tanith: Most of us won't know anything about it.

Diane: It'll be kept hidden from us? Again?

Tanith: It'll be too scary for most people to handle, so most people won't really know. It will be what leads us into that leap of technology, technological change.

Diane: So that technology will come from other planets?

Tanith: Yeah, the start of it, the kick of it will. Since time doesn't really exist, some other planets are just future places, and they can even be future places of this planet. As that comes back to us we'd get that kick, somebody gets that contact and gets that kick of technology. But we won't know, many of us won't know. Many of us will wonder. It isn't that the government's going to hide it from us, they're going to hide it from themselves 'cause they won't know what to do.

Diane: That's been going on for a long time.

Tanith: Yeah, and it's just 'cause it'll be too scary. And actually that's good because they need, we need, the changes to go through more, so there is no Pentagon to destroy anything good that comes, and try to dissect it and figure out what it is.

Diane: Will that happen? That there'll be no Pentagon?

Tanith: Yeah. It'll be there, it'll be a big building with a bunch of good old boys who think they're important and no money to support it. They'll have some things. They'll have a National Guard to come help the people after a flood; they'll have a navy to watch the shores. More of their job will be pollution watch, animal watch — it'll be a watch. It'll be there because the men who have moved up in those ranks need somewhere to go, and so we'll give them a ship and tell them to patrol the shores and make sure we're safe. They'll all feel very proud of themselves, and they'll leave us alone. It's like giving a child a toy and saying please play with this and leave me alone, and so we'll do that.

Diane: Only don't shoot it at anybody, this toy of yours!

Tanith: Yes. We don't have any money for bullets but you can have the guns.

Diane: Hmm. Sounds like an improvement.

Tanith: Yeah, but it's going to be like a weaning process for them because it's not going to just end. But it'll be okay because...

There's like a pocket in the earth forming that's being filled with the energy of the Great Mother. She's pulling all of her energy to one great pocket. Two great pockets. One is in the ocean, under the ocean, and that's a safety pocket. That's a pocket of all the spirit essence of the whale, the spirit essence of the dolphin, the spirit essence of the hawk, and the eagle, and the elephant, and the whole endangered list. The spirit essences are being saved. It's like Mother Earth's ark, and it's under the ocean. It's a pocket, it's very safe, if we all die she will create those again.

The other pocket is where she's putting all of her essence. Her essence has always been all around the world. She's pulling it in, she's pulling it into a pocket, and in that pocket will be her center of energy and her ability to make the changes, to really push the changes that need to happen on this planet. She has a place where she's saved in safety all that she might want to recreate and a place for herself. Because at this time she is becoming the Dark Isis and the Dark Kali, and the black side — which isn't evil but it's the powerful side. It's the powerful side of that woman/Mother that says, I can kill to save. And I am also death at the end of all things, even though I am life at the beginning of all things. She is becoming that essence to move these changes, to move them, and so she is pulling herself in. Where she is leaving behind is where we're seeing great desolation, and that is one of the ways she's doing the changing. By leaving, pulling her spirit from a piece of land, from a place, and saying you don't have me now and do what you must. What's happening in those areas is the desolation and the destruction.

Diane: How is that for individuals? Is she doing that with individuals also?

Tanith: The ones that will survive are already picked. I mean they can still earn it, others can still earn it, but if you're in an area of utter destruction there's a good chance you're going to be destroyed. But many of those who need to survive simply won't be there. They'll simply be gone; they'll be on vacation. They simply won't be there when that destruction hits or for some reason they will be saved. That's part of the great changes happening in people right now, and more so in women because they are so tied to the earth. So we feel that great shift in the Mother energy, and that the Mother energy is now becoming that dark energy. It's almost like you've been a two-year-old and your mother's really lost her temper now. That's part of what mothers have to do is once in awhile tell you that you've really gone out of line. And that's about what she's going to do. She hasn't lost her loving warm aspect, and this isn't an evil, nasty, cruel, punishing aspect. It's an aspect that says, I will not let you destroy everything. I will pull to safety what needs to come to safety, and let those of you who cannot make the changes and cannot understand live on the land that does not have my energy in it. And that land will not survive, because it doesn't have her energy in it.

There's still that good side, there's still that loving side — and it's not good or bad — there's still that Maiden side and the Mother side and the mothering, nurturing side, and that will always be there through everything. That will still be there for us to tap. But part of what is being done is that the death side — the death aspect, the Crone aspect — is taking over a little bit and saying, uh uh. Anything that does go extinct she has that chance to reintroduce, and it'll be one of those miracles like we've already seen, where we thought something was extinct and then we found three or four somewhere, where they shouldn't have been. She's been doing it for awhile. She has that capability.

Diane: It's a comforting thought right there.

Tanith: Yeah. That's what's important to her. That's what's close to her heart, and that's why animals will become so much more important because they'll be there.

Diane: Why haven't they been protected all along? Is it twenty species a week that are becoming extinct?

Tanith: She's been very hurt and very upset, and she's been in her Mother aspect and hasn't turned into that warrioress aspect that says, Let's have no more shit, Quemosabe. It's sort of like that. She hasn't been there. She's been the nurturer, the Mother, the lover, the forgiver, and that's why she knows now she needs to be more than that. She needs to also be the Crone, to be the one who draws the line and says, no more.

Diane: How long for this process?

Tanith: The animals come back over a very long time, but definitely there's a twenty-thirty year something here where there's a big change, there's a big social change. The whole natural order changes.

Diane: How long after the actual earth changes until culture is reestablished again in a better way?

Tanith: Well, see it isn't totally going to fall apart. It won't ever be totally fallen apart; it'll be a big process of transition. We're not going to go into the dark ages and come back out. It's going to be just a lessening on all fronts. Instead of people listening to one thing, they'll be listening to the other. Time for the earth planet is very different from time for us, so a hundred years for her is nothing. A hundred years for her is not even a blink of the eye.

Diane: But things are speeding up real fast right now.

Tanith: Yeah, because she's angry. She's a little bit angry.

Diane: How about the process in individuals? Individuals are going through such major major stuff now.

Tanith: It's a test. It's a test to see who can look at their issues. Who can look at what's really really happening and what they really do, and get beyond it and move forward into the next stage. Right now her concern has shifted away from helping us all through to saving the planet, so it's almost like we have been children for a very long time. The Mother's always been there for us, and now Mother is saying, Look I'm really busy and you need to go take care of this yourself. And for the first time we are not without her, but more on our selves. When you call on her she will listen, but you need to call. She's doing other things right now and her focus is somewhere else. She remember who calls, and it's really good if you call. I mean, if you call on her she will listen, but you need to call. She's doing other things right now and her focus is somewhere else. She remembers who calls, and it's really good if you call. I mean, if you call it's not against you, it's an acknowledgement of, I know you're there for me still.

But it's happening to people because she can't do people and animals and planet and everything all at the same. It's gotten out of hand and she needs to focus. And because her aspect is more the Crone aspect, which is the wisdom aspect, we are all now heading very quickly from that nurturing aspect to that Crone aspect. We're not aging physically, but we're aging maturity-wise and we go through it. Aging is a growth process and it's a painful process and so we're going through it quicker. We're going through this period of her bringing out her Crone energy and the earth having its Crone energy. After that we'll go back to the Maiden. But we have to go through the Crone, and the Crone can be scary.

For some people it's like the curtain: a few days ago I saw what was on the other side of the dark curtain, and all it was was light. It's the dark curtain that is the Crone and it's the light that is the Maiden. And it is simply the movement through the Crone that lets us grow and mature and be who we need to be, and be reborn as the Maiden. And men do the same things, some — if they can get through the Crone — if they can get through that. Men are terrified of the Crone, and they can't get through that. Many of them will either send off to play somewhere else or will not go through the Crone and will not be with us.

Diane: Wars?

Tanith: Little wars, civil wars, stupid wars. Mothers won't be sending their sons anymore, it'll be the sons who want to go. There will not be another war of, "your mother is proud to see you die." It will be terrible and it will be necessary and it will be fine, because that is what those people need to do.

There'll be a danger for women of becoming avatars, an ego danger for women and we're going to see that intensifying, much like the dangers right now inherent in Craft for high priestesses where they get too egotistical. There's going to be a lot of people listening to women because they will be the ones transmitting this message and have transmuted that energy, and somehow people just feel good around them. Somehow their lives come out of the morass and become something beautiful and wonderful like a phoenix rising again. People will be attracted to them, and there will be a danger that'll be one of the other tests for these women. It's alright to have some ego about it because that's okay, and it's alright to even live by this, if you are able to support yourself by this, because that's okay. It's *not* okay if you do not become one that is connected with the others, with the other sisters. If you are not connected to the link of the sisters, you become only yourself and recreate that whole problem again. There will be a danger there for women.

Diane: So there needs to be a network.

Tanith: Yeah. The women who are networking to network, that's not where the right connections are being made. That's okay to do, but the right connections are almost by accident. They're just sort of happening, kind of like you and Gael, kind of like you and me, kind of like — they're just kind of happening. But if I were to write to Selena and try to network with Selena and try to establish that, it wouldn't be an important contact. The ones that happen, that just happen, will be the ones that need it and will weave the link that is to be established.

Diane: It's always been that way.

Tanith: And they may never know it; they may never ever know it. They may never know that they're part of a psychic link that is connecting or the importance of it. It may not be necessary for them to know.

Diane: I think aware women do know.

Tanith: Yeah, but not yet. And it's not necessary, it doesn't really matter. It's okay if they don't know. 'Cause one of those women might be a little old lady in Craig, Colorado, and doesn't know anybody, but has two friends. One's in California and one's in New York, or one's in Montana, and she's part of that network somehow. She doesn't know it. She just knows that it's important to listen here and send it there. She has no idea why, and that's okay. It isn't important for her to know; it's important for her to be part of that connection, that network that creates that other energy. It isn't important for the Goddess religion to grow and become masses and masses of people who do three thousand-people rituals in Central Park. It isn't important; it just isn't important.

Diane: No, I think that would hurt it.

Tanith: Yeah, it's just not important. It's always going to be that quiet, quiet Goddess worship that is done in houses and back yards and wherever. Some people are very close to the right track, the Goddess sanctuaries are very close and will become a central point.

Diane: Will these become living communities?

Tanith: Some women might, some women might choose to do that, but that isn't the way of the world. But I think some women will choose to do that.

Diane: A lot are starting to talk about it now.

Tanith: I think some women will choose to do that, and I think you'll find that many of the women who were abused, gay or straight, the women who have really really suffered....The women who know they have a role but don't feel they can do it by themselves because they're still under that suffering, will come together and create a community where there is a school. Now, we're going to see more of that anyway, because there'll be community schools and community....Not total, not total socialistic, but there will be more community kinds of things happening. But I think there will be women who realize that part of the abuse was the healing process that was put on them by a patriarchal doctor. And now if you were abused a woman says to you: You can have your virginity back, we can do a ritual and give you your virginity back, we can do a ritual that says the pain is gone, it's released. If they're not told that and can't do that, they always are the victim because they haven't ever simply released it, turned around and said, I don't need to relive this.

Diane: And that's the kind of thing that's so terribly hard for women right now.

Tanith: So those women will be finding that they work better in a women's community where they can totally separate themselves. I don't think there'll be the man-hatred as much.

Diane: No, that's declining.

Tanith: The men will be changing, too. We'll be getting rid of those men who are worth hating. And what we'll be finding is that some of those women's communities will be turning into healing communities.

Diane: So there will be healing communities?

Tanith: Yeah, and I think that will be a choice for some people, instead of going to what's left of doctors and drugs.

Diane: Yeah, I can see myself as part of something like that.

Tanith: You could probably run one, because part of what will happen, too, is many of these women are not good at running things because they've been taught for too long they can't. There'll be a very few women who are able to seed one and run it, and that'll be very important, too.

Diane: And healing will find its place over the medical system, then?

Tanith: Yeah, but not publicly. I mean it will because people will just do it that way. But it won't be that doctors, trained medical doctors, start accepting it.

Diane: I know some trained medical doctors that are wonderful healers. I'd like to see more and more of them.

Tanith: There might be, there'll be some, but it isn't like the whole medical practice will do that. It's just that this will become such a big force it can't be ignored anymore. People will simply go there instead of going to a doctor.

Diane: Without the healers being persecuted?

Tanith: Yeah, because they'll have banded together in these communities. They will have their own protection.

Diane: 'Bout time.

Tanith: Yeah, it's about time for the people, too. It's about time for the people to become aware. It's a ways down, you know, it's still a little ways, it's not tomorrow.

Diane: How about psychic skills. You talked about telepathy. I have a vision of seeing healing and telepathy and trancework taught.

Tanith: Yeah, there'll be telepathy.

Diane: I have a vision of seeing these things to be common mainstream knowledge, taught universally in the gradeschools.

Tanith: Well, I don't know whether it's going to be taught; I think it's just going to be acknowledged. 'Cause I don't know that psychic stuff is taught. I think it's there, you're born with it. What's taught now is to not use it. If it's just not taught away then it's still there.

Diane: Right.

Tanith: And I think we're going to be finding the strain of it is just a little stronger. It's been dying out somewhat in children in the same way that creativity has, and trained away, and there are many women who have the power not bearing children. But what you'll find is as we let that creativity come back and that acceptance is just there, it's not taught away, then that will come back. And the women who are the connectors will take students — many of these women who are the connectors will not have their own children — but will be teachers.

Diane: And we'll also see that happening, too?

Tanith: Right, and they are supposed to not have children.

Diane: Okay. But that's not a loss to the culture?

Tanith: Uh, no. It was a personal sacrifice for many of them. Many of them thought they never wanted kids but on some level it really was a personal sacrifice. Because there are many who need the training and they could not have the children, have the psychic sense, and be the teachers. They couldn't do all of that. Many times when you're pregnant and you give that much creativity into another human being, it's not that you lose it but you go into a time where you can't use your abilities as much in other realms. It's going into creating that human. It's much like a woman who's a doctor but when she's pregnant she's not quite as good at her practice. Not that she's not a good doctor, but the practice doesn't have as many clients. She doesn't have the energy to put into it, and it's not tired energy, it's creation energy that's going into the baby. So for many of these women, they need to not have children.

But they *will* have children. They will have children who knock on their door, and children who come, and children who are foster children, and they will be in the end fulfilled and have those children. They may get them when they are sixteen or seventeen or twenty; they may be fifty and get a twenty-year-old, but that bond will be there and it will be their child. And that training will be there, that teaching will be there.

Diane: It doesn't matter that the heredity is being lost? These are the women with the psychic skills.

Tanith: It's not being lost, the Mother has them. It's in that pool, it's in that sacred pocket, it's in there. She can bring it back. She can bring it back to us. It's because they just have other things they need to do. They need to put that into the earth, they need to put that into the world around them. They need to do these other things.

Diane: And these women reincarnate anyway, and will do it then.

Tanith: Maybe not here.

Diane: Wherever.

Tanith: Yeah, but maybe not here. Maybe and maybe not.

Diane: Where else would they go? Where else would they choose to go?

Tanith: I don't know. There's another — and see I don't know if it's another planet — I think we talk about planets because it's beyond our comprehension to think in dimensions and in time.

Diane: Yeah, more and more are talking about dimensions.

Tanith: It's beyond our comprehension. Three dimensions we can get, four dimensions is a little rough.

Diane: We're moving into the fourth now.

Tanith: Five, six, ten is really hard. For us to think about other spheres that exist somewhere, but maybe not somewhere we can see through a telescope, would be very very difficult for us.

Diane: Mari talked about Atlantis in that light.

Tanith: I don't think Atlantis was on this planet.

Diane: She feels it was a different dimension.

Tanith: Okay. It wasn't in this physical.

Diane: Right.

Tanith: And especially 'cause geologically I can't find it. Geologically it wasn't there. Geologically there was *something* that went underground that was flooded, but not the stories of Atlantis. And the stories of Atlantis are too prevalent to never ever have existed.

Diane: Right.

Tanith: But what did exist was not what we are capable of comprehending, okay? In fact the story of the evil that was in Atlantis and destroyed it, I don't think that's right either, because I think it was a time period of the black, of the Dark Isis. We can't comprehend that. It wasn't evil; it was the Crone aspect, only deeper than the Crone, 'cause you know the Dark Isis is before the Isis. It's that deep deep, incredibly powerful realm.

Here's my other side kicking in, my other side is saying it just got this message that's so bizarre, *I'm* not going to say it! Somebody might read this book, and they're going to think I'm crazy! Women are going to have the option of developing wings again!

Diane: Oh, that sounds wonderful!

Tanith: I think I should write about that to AmyLee because she said something about the Indians know that women had wings. Women will be able to very very slowly, very slowly, over time find their wings again, develop their wings.

Diane: Is this teleportation, or is this physical wings?

Tanith: I'm getting physical wings.

Diane: Gosh, I want to see that! I want some.

Tanith: I'm getting wings. Your body could have something attached to it that's not seen in this dimension, and those wings could be in that other dimension. They're really attached to you, but no one gets to see them, kind of like auric wings or etheric wings — you know, that other-part-of-you wings. Women will have the chance to get those back, if we want them.

Diane: Ah, neat!

Tanith: I think I'm first on the signup sheet, I'm not sure.

Diane: If you're first, I'm second!

Tanith: That'll be a gift, that'll be a gift for those women. And you know there are a few men, it's mostly women, but there are a few men.

Diane: Could we have tails to wag, too? (laughs)

Tanith: Probably, if you want them! Maybe men get tails, I don't know.

Diane: (laughs)

Tanith: Women get wings and men get tails? Something to chase when they're bored.

Diane: Onyx is looking at me like, did you have to?

Tanith: What do you think, Onyx, you want to say anything else? Onyx? Is there anything else to say? Puppies rule. Dogs are important.

Diane: Yes, dogs are very important.

Tanith: Dogs have lessons to teach us that we have been so stupid to never learn.

Diane: Well, some of us aren't stupid. Don't write off the whole race, puppy.

Tanith: There'll be some men, too, in that network and in everything. There will be some men. There will be groups of women who are just groups of women, but it isn't like it's all women. There are some — very few but there are some — men who will be part of that psychic network, part of that teaching network. The ones who understand what their true role is, and what they really are.

Diane: There are beginning to be some now.

Tanith: Yeah, the protectors. The ones who understand that they are the protectors, they're beginning to be there. As they continue, they will spread that message among other men and that segment will be able to mutate and grow. Not that men will mutate, but how they've been using that energy and stuff, that will mutate.

Diane: They'll learn creative ways instead of destructive ways.

Tanith: Right.

Diane: Will the diseases be gone eventually?

Tanith: Not all of them; people still have to die.

Diane: They don't have to do it in such horrible ways, though.

Tanith: Well, the fear of dis-ease will go. When the fear of something goes then the horribleness about it goes. People can die of cancer and have a very nice death, have a very nice, quiet peaceful death. If they know that. If the fear of the dis-ease goes, the fear of death goes. When someone's dying, we maybe need to celebrate that death, and understand that.

Diane: Will that be more of a universal knowledge?

Tanith: That'll be more of a cultural knowledge. Yeah, yeah it will be, because the places where they fear the dis-eases less will have less dis-eases. Places where you fear it more and have more dis-eases, there'll be more death in those places. It'll still be there.

Diane: And death will be more accepted as a natural thing and as part of the process?

Tanith: It'll be easier. I think death is going to be easier. The goal won't be quite so much keeping them alive.

Diane: There's such tremendous denial in this culture about death.

Tanith: I think that will fade some because it'll almost be a numbing effect of how much death there was on the planet.

Diane: How about an acceptance or awareness of reincarnation?

Tanith: For some.

Diane: For only some? Not universally?

Tanith: It isn't necessary universally. It isn't necessary, as long as they don't punish others who believe it. Isn't necessary, because it's not that everyone on the planet will be an old soul. Some might be young souls and not understand yet and not be ready for the knowledge. And it's alright.

Diane: Okay. Anything else?

Tanith: Just that there's a lot of hard times but not the incredible destruction that we've always feared, and after there can be a lot of quiet joy, much joy and celebration. It isn't going to be us living in sorrow, but we will be taught that we are simply part of this whole system and cycle, the living, breathing earth planet we live on. We are simply part, and we simply have to settle for that — being part. Not the rulers, not the best, not the smartest, not the wisest, but simply part.

Diane: Cooperate?

Tanith: Yes.

Diane: Communal?

Tanith: For some it'll be communal, for some it won't. For some there's gonna be those that live on their property in the middle of their eighty acres in the middle of nowhere, and they don't want to be communal. And that's okay. It isn't that all people are going to be the same. We'll still have personality differences and we'll still have people who want different things, and who like different things, but there'll be that difference. There'll be the difference in that we're aware that we're only part of a great whole.

Diane: That we're not the only species, or the only species that counts.

Tanith: We're like part of a breeze in a very very large...Right now we're part of a breeze in a tornado, but we started the tornado. We have to learn that we can't control the tornado, that we don't get to, and it's better to not have them. It's better to be satisfied just being a little breeze by yourself without having to be a tornado to be impressive.

Diane: This is a demanding puppy. Everytime I stop scratching her chest I get whacked. (laughs)

Tanith: I know! "I'm not going to settle for anything less!"

Diane: She's also been involved with this whole thing. She hasn't gone to sleep once.

Tanith: I know. She keeps saying, dogs are important, dogs are important.

Diane: I *know* that.

Tanith: Yes, dogs are important. Dogs are important. I know, I know. Get that message on tape, okay? One track mind.

Diane: Chewees are important, too.

Tanith: Yeah. I don't know; I guess we're done?

Diane: Sounds like it. Thank you.

Notes

1. This is not racial segregation, but separation by awareness and the ability to live in peace.

Channelers

Mari Aleva
Moonwind
P.O. Box 85986
Westland, MI 48185
(313) 326-7561

Laurel Steinhice
The Crystal Circle
P.O. Box 50145
Nashville, TN 37205
(615) 356-4280

Tanith
P.O. Box 280-341
Lakewood, CO 80228

Marion Webb-Former
24 Roberta Lane
Lowell, MA 01852

Bibliography

Aleva, Mari. *Voices From Beyond*. Westland, MI: Moonwind Publications, 1991.

Arguelles, José. *The Mayan Factor: Path Beyond Technology*. Santa Fe: Bear and Company, 1987.

Bryant, Page. *The Earth Changes Survival Handbook*. Santa Fe: Sun Books, 1983.

Cayce, Edgar Evans. *Edgar Cayce on Atlantis*. New York: Warner Books, 1968.

Chaney, Earlyne. *Revelations of Things to Come*. Upland, CA: Astara Inc., 1982.

Charnas, Suzy McKee. *Motherlines*. New York: Berkeley Books, 1978.

Childress, David Hatcher. *Lost Cities of Ancient Lemuria and the Pacific*. Stelle, IL: Adventures Unlimited Press, 1988.

Churchward, Col. James. *The Lost Continent of Mu*. Albuquerque: BE Books, 1959.

Clow, Barbara Hand. *Eye of the Centaur*. St. Paul: Llewellyn Publications, 1986.

Gadon, Elinor W. *The Once and Future Goddess*. San Francisco: Harper and Row Publishers, 1989.

Gearhart, Sally Miller. *The Wanderground: Stories of the Hill Women*. Watertown, MA: Persephone Press, 1978.

Jochmans, J.R. *Rolling Thunder, The Coming Earth Changes*. Santa Fe: Sun Books, 1980.

Johnson, Buffy. *Lady of the Beasts: Ancient Images of the Goddess and Her Sacred Animals*. San Francisco: Harper and Row Publishers, 1988.

Petersen-Lowary, Sheila. *The 5th Dimension: Channels to a New Reality*. New York: Simon Schuster, 1988.

Montgomery, Ruth. *Threshold to Tomorrow*. New York: Fawcett Books, 1982.

Montgomery, Ruth. *The World Before*. New York: Fawcett Books, 1976.

Piercy, Marge. *Woman on the Edge of Time*. New York: Fawcett Books, 1976.

Summer Rain, Mary. *Phoenix Rising: No-Eyes' Vision of the Changes to Come*. Norfolk: The Donning Company, 1987.

Summer Rain, Mary. *Spirit Song: The Visionary Wisdom of No-Eyes*. Norfolk: The Donning Company, 1985.

Robinson, Lytle. *Edgar Cayce's Story of the Origin and Destiny of Man*. New York: Berkeley Medallion Books, 1972.

Sargent, Pamela. *Venus of Shadows*. New York: Bantam Books, 1990.

Sitchin, Zecharia. *The 12th Planet*. New York: Avon Books, 1976.

Sjöö, Monica and Mor, Barbara. *The Great Cosmic Mother: Rediscovering the Religion of the Earth*. San Francisco: Harper and Row Publishers, 1987.

Snow, Chet. *Mass Dreams of the Future*. New York: McGraw-Hill Publishing Co., 1989.

Starhawk. *Truth of Dare: Encounters with Power, Authority and Mystery*. San Francisco: Harper and Row Publishers, 1987.

Stein, Diane. *Casting the Circle: A Women's Book of Ritual*. Freedom, CA: The Crossing Press, 1990.

Stone, Merlin. *Ancient Mirrors of Womanhood: A Treasury of Goddess and Heroine Lore from Around the World*. Boston: Beacon Press, 1979.

Stone, Merlin. *When God Was A Woman*. New York: Harcourt Brace Jovanovich, 1976.

Sunlight. *Womonseed*. Little River, CA: Tough Dove Books, 1986.

Temple, Robert K. G. *The Sirius Mystery*. Rochester, VT: Destiny Books, 1976.

Tepper, Sherri S. *The Gate to Women's Country*. New York: Bantam Books, 1989.

Valentine, Ann and Essene, Virginia. *Cosmic Revelation*. Santa Clara, CA: S.E.E. Publications, 1987.

Waters, Frank. *Book of the Hopi*. New York: Ballantine Books, 1963.

Gonzalez-Wippler, Migene. *Tales of the Orishas*. New York: Original Publications, 1985.

Young, Meredith Lady. *Agartha: A Journey to the Stars*. Walpole, NH: Stillpoint Publishing, 1984.

Also by Diane Stein

Casting the Circle
A Women's Book of Ritual

All Women Are Healers
A Comprehensive Guide to Natural Healing

The Goddess Celebrates
An Anthology of Women's Rituals

❋ *The Crossing Press*
publishes a full selection of women's
spirituality, healing and feminist titles.
To receive our current catalog,
please call—Toll Free—800/777-1048.